"Construction costs are a never-ending subject of conversation between scholars and practitioners in the construction domain. This book offers definitive guide to the holistic dimensions of construction cost management for academics and postgraduate students. In particular, the insight into research strategies and decision-making tools adds value to existing knowledge-gap in this area."

- Che Khairil Izam Che Ibrahim, Associate Professor, Faculty of Civil Engineering, Universiti Teknologi MARA, Malaysia.

Continuous Cost Improvement in Construction

Continuous Cost Improvement in Construction: Theory and Practice aims to provide students and practitioners with an all-inclusive understanding of strategies for adopting continuous improvement in construction cost management. This book addresses continuous improvement practices from the perspective of cost management and applies case study examples to question the readers' perspective of continuous cost improvement strategies in the project lifecycle. Continuous cost improvement practices in managing the cost of minor, major, and mega projects are all connected with decision-making tools for devising strategies for choosing the approaches for mitigating the effect of cost overruns in construction projects. Continuous cost improvement should be taught as part of modern methods and processes of construction in further and higher education institutions. This book will be key reading for all advanced undergraduate and postgraduate courses in Construction Project Management, Building and Quantity Surveying. Professionals in all aspects of the AEC industry will also gain greatly from engaging with the key concepts of continuous cost improvement throughout this book.

Temitope Omotayo is a Senior Lecturer in Quantity Surveying at Leeds Beckett University, UK. He holds a PhD in Construction Project Management from the University of Salford, UK after gaining an MSc in Quantity Surveying (Mechanical and Electrical) from the same University. Temitope is a fellow of the higher education academy (FHEA) and holds a postgraduate certificate in higher education (PGCAP). His research interests are continuous improvement in construction, the application of systems thinking for decision making in the built environment, sustainable construction, and smart cities.

Udayangani Kulatunga is a Professor in Building Economics, attached to the Department of Building Economics, University of Moratuwa in Sri Lanka. She is also the Director of the Centre for Disaster Risk Reduction at University of Moratuwa, Sri Lanka. Udayangani completed her undergraduate degree in Quantity Surveying (Hons) at University of Moratuwa, Sri Lanka and her PhD at University of Salford, UK in Performance Measurement in construction research and development. Udayangani has teaching experience both in the UK and in Sri Lanka. She is a fellow of the Higher Education Academy (FHEA), UK and holds a postgraduate certificate in higher education (PGCAP) as well.

Bankole Awuzie is an Associate Professor, Department of Built Environment, Faculty of Engineering, Built Environment and Information Technology at the Central University of Technology, Free State, South Africa. He holds a PhD in Built Environment (Construction Project Management) from the University of Salford and an MSc in Construction Project Management from the Robert Gordon University, Aberdeen, Scotland, both of which are in the UK. He is a rated researcher of the South African National Research Foundation (NRF). His research interests span the smart, sustainable, and circular construction knowledge domain.

Continuous Cost Improvement in Construction

Theory and Practice

Temitope Omotayo, Udayangani Kulatunga, and Bankole Awuzie

Routledge
Taylor & Francis Group

LONDON AND NEW YORK

First published 2022
by Routledge
2 Park Square, Milton Park, Abingdon, Oxon OX14 4RN

and by Routledge
605 Third Avenue, New York, NY 10158

Routledge is an imprint of the Taylor & Francis Group, an informa business

British Library Cataloguing-in-Publication Data
A catalogue record for this book is available from the British Library

Library of Congress Cataloging-in-Publication Data
Names: Omotayo, Temitope, author. | Kulatunga, Udayangani, author. |
Awuzie, Bankole Osita, author.
Title: Continuous cost improvement in construction : theory
and practice / Temitope Omotayo, Udayangani Kulatunga,
Bankole Awuzie.
Description: Abingdon, Oxon ; New York, NY : Routledge, 2022. |
Includes bibliographical references and index.
Identifiers: LCCN 2021047552 (print) | LCCN 2021047553 (ebook) |
ISBN 9781032008448 (hbk) | ISBN 9780367774585 (pbk) |
ISBN 9781003176077 (ebk)
Subjects: LCSH: Building–Cost effectiveness. |
Continuous improvement process.
Classification: LCC TH438.15 .O46 2022 (print) |
LCC TH438.15 (ebook) | DDC 624.068/1–dc23/eng/20211123
LC record available at https://lccn.loc.gov/2021047552
LC ebook record available at https://lccn.loc.gov/2021047553

ISBN: 978-1-032-00844-8 (hbk)
ISBN: 978-0-367-77458-5 (pbk)
ISBN: 978-1-003-17607-7 (ebk)

DOI: 10.1201/9781003176077

Typeset in Baskerville
by KnowledgeWorks Global Ltd.

Contents

Figures

Tables

Foreword

Effective cost management is a major requirement for successful completion of construction projects. That said, cost overrun is a major problem in project development and a regular feature in the construction industry, which makes project costly for the parties involved in construction.

The construction industry must learn from other industries and adapt to changes faster to mitigate the challenges of cost overrun, low quality, contractual dispute, poor decision-making, and the rising cost of construction. The manufacturing sector offers some practices that can help the construction industry. One of the strategies applied by this sector to reduce production costs and maximise profit is borne out of the lean thinking philosophy with the associated continuous improvement which provides a strategic benefit of eliminating non-value adding activities in the production process for cost reduction.

Continuous cost improvement in construction: Theory and practice therefore introduces a new pathway for construction cost management. The book presents new strategies, cases, and decision tools as alternatives for improving construction cost and decision-making in construction management. Continuous cost improvement is not just a theoretical knowledge but a collection of opportunities for solving construction cost management problems.

The book raises a wide range of issues relating to construction cost improvement underpinned by construction cost management that seek to drawn alignment across a wide range of players involved in construction development, as they have significant bearing on the manner in which cost innovation activities occur in the industry.

The authors have shown how and why traditional cost management, system, methods and techniques are limited to deal with the issues of cost overrun and managerial decision alternatives and particularly the practice for modern methods of construction. In this new book, the authors show how continuous cost improvement is a key to address this and present superior decision-making tool for competitive advantage. This book therefore seeks to shed light on the way forward for the construction industry in the age of globalisation of knowledge across diverse industries and the role cost innovation plays in determining the position of the competitiveness of construction firm.

Overall, this book offers a useful and in-depth look at ways to design and implement procurement strategies, contractual payments, modern methods of cost control, cost reduction and maintenance through continuous cost improvement practices. This is clearly written, well organised, and will be of enormously practical value to academics, researchers, and practitioners who are involved in cost management and improvement of construction projects.

Professor Akintola Akintoye
Dean, School of Built Environment,
Engineering and Computing,
Leeds Beckett University

Preface

Continuous improvement focusses on incorporating incremental changes to the activities performed in an organisation, to deliver value for all stakeholders. Having its roots in the manufacturing industry, continuous improvement draws interest in the construction industry too. Within this context, continuous improvement of cost has been able to attract more attention, as "cost" has been considered as the vital component of any construction project.

The aim of this book is to provide students and academics with an all-inclusive understanding of strategies for adopting continuous improvement in construction cost management. A variety of theoretical and practical knowledge concerning continuous cost improvement in developing and developed nations, decision making tools that can be used for continuous cost improvement and applications of cost reduction mechanisms in minor, major and mega projects around the world are presented in this book. Hence, the combination of differing perspectives on continuous improvement in construction management from research, organisations, project and around the world with examples have been presented. The overall outcome of reading this book describes and illustrates strategies for choosing the right continuous improvement approaches for mitigating the effect of cost overruns in construction projects. Despite the general conception that reduction of cost would ultimately reduce the quality, this book suggests otherwise, that continuous cost improvement would improve the quality, employee-employer relationship, client satisfaction and the overall growth of the organisation whilst serving as a panacea for challenges resulting in cost overrun.

The importance of eliminating waste, improving productivity, and reducing costs are key considerations of construction industry at all times. Yet, during the recessions and in pandemics such as today, the importance of continuous cost improvement in construction is ever increasing in reducing the costs. Hence, theories and practicalities of continuous improvement in construction cost management would be essential to transform the industry towards its "next normal" as well.

Having worked together in many scholarly works in academia, related to construction and Quantity Surveying education in particular, the authors got together from three continents, Europe, Africa and Asia to compile this book during the most challenging and difficult times of the COVID-19 pandemic. The authors presume that the readers would gather useful theoretical and practical knowledge on continuous cost improvement in construction from this book.

Temitope Omotayo
Bankole Awuzie
Udayangani Kulatunga

Acknowledgements

We would like to express our gratitude to all those who have contributed information and advice leading to the completion of this book. We also express our special thanks to our families have also been supportive throughout the intense months of writing during COVID-19 restrictions.

SECTION A
BACKGROUND

1 Continuous Improvement and the Construction Industry

Temitope Omotayo, Udayangani Kulatunga, and Bankole Awuzie

1.1 Introduction

Between the late 1800s and early 1990s, attention was given to management practices geared towards alternative means of solving production problems and the development of new labour standards. In the process of alleviating production and labour challenges, the United States (US) government established the "Training Within Industry" service during the Second World War. Training Within Industry comprised job method training and programmes designed to train and educate supervisors on the benefits of continuous improvement in the workplace.

On the international stage, concepts and application of continuous improvement were introduced in Japan after the Second World War (Kapur, Adair, O'Brien, Naparstek, Cangelosi, Zuvic, Sherin, Meier, Meier, Bloom & Potters, 2016; Maarof & Mahmud, 2016; Suárez-Barraza, Ramis-Pujol & Kerbache, 2011). In Japan, continuous improvement is known as *Kaizen*, which literarily means "change for the better" (Omotayo, Awuzie, Egbelakin, Obi & Ogunnusi, 2020a; Singh & Singh, 2015). The Toyota Production System in Japan was one of the first organisations to implement continuous improvement in their production line. Shortly after the Second World War, the US wanted to encourage Japan to rebuild their country. Gen. MacArthur tasked developmental experts in the US to visit Japan for the sole purpose of helping the Japanese to rebuild. Dr Edward Deming, a statistician, was one of the experts who explored a new mechanism for creating a manufacturing workforce in Japan. Prior to the infrastructure issues plaguing Japan in the 1940s, continuous improvement had already been a culture of Japan. However, in the past 30 years, continuous improvement has transformed into a major part of the lean philosophy. Continuous improvement has not only been used in the manufacturing sector but also in other sectors, such as health care, agriculture, sport, governance, quality management, human resource management, and construction (Imai, 1992; Kapur *et al.*, 2016; Prashar, 2014; Suárez-Barraza *et al.*, 2009; Suárez-Barraza & Ramis-Pujol, 2010). There are many construction industry practices that are synonymous with

DOI: 10.1201/9781003176077-2

the concept of continuous improvement. Target costing, waste reduction, and post-project reviews are being practised in many construction companies globally.

In the construction industry, continuous improvement in managing construction cost is a relatively new concept. Continuous improvement in cost management has been suggested as a panacea for the challenges of cost overrun. Many authors, such as Koskela, Ballard, Howell and Tommelein (2002), Macomber, Howell and Barberio (2007), and Omotayo *et al.* (2020a) viewed construction cost management from the perspective of lean construction, target value design, and Kaizen costing as an opportunity for managing construction cost. Nevertheless, there is a dearth of literature addressing continuous improvement in construction cost management. For a clearer understanding of how continuous improvement can work with construction cost management, there is a need to highlight the definition and concepts of continuous improvement and construction cost management.

1.1.1 Defining continuous improvement

The definition of continuous improvement has evolved in the last 40 years. Although the general opinion concerning continuous improvement emerged from the Japanese word known as "*Kaizen*", the "change for better" mantra has blended into the various sectors of its application. Sanchez and Blanco (2014) studied the various definition of continuous improvement in the last 40 years. The following definitions were extracted from these studies.

Deming (1982. p. 2) described continuous improvement as "*improving constantly and forever the system of production and service*". Deming's description of continuous improvement was from the perspective of the production system and human resource management.

Imai (1992, p. 71) defined continuous improvement as "*finding a better way of doing the job and improving the existing standard*". Imai's explanation of continuous improvement addresses the non-ending training of all talents in organisations as a tool for enhancing organisational productivity and growth.

According to Bessant, Caffyn, Gilbert, Harding and Webb (1994, p. 26), "*a company-wide process of focused and continuous incremental innovation*" is the most effective definition of continuous improvement in an organisation.

Juergensen's (2000, p. 3) definition of continuous improvement is very generic and is applicable to any sector, namely "*improvement initiatives that increase successes and reduce failures*".

Bessant, Caffyn and Gallagher (2001, p. 76) defined the concept as "*a particular bundle of routines which can help an organisation improve what it currently does*". Their perspective is based on organisational growth and activity enhancement.

Dahlgaard, Kristensen and Kanji (2010, p. 445) mentioned: *"small continuous changes for the better"*. Furthermore, Dahlgaard et al. (2010) provided a generic micro-level understanding of how continuous improvement can be achieved in an organisation, in methods, and in processes. Small changes in organisations can be incremental and continuous.

Brunet and New's (2003, p. 1428) definition, namely *"pervasive and continual activities, outside the contributor's explicit contractual roles, to identify and achieve outcomes he believes contribute to the organisational goals"* includes change management in the process of continuous improvement. The authors' definition also understands the strain of continuous improvement on the employees within an organisation and thus acknowledges the role of a continuous improvement champion.

Boer and Gertsen (2003, p. 882) viewed continuous improvement from the perspective of activities carried out within a system. For continuous improvement to fit into any system, effective planning is vital: *"The planned, organised and systematic process of ongoing, incremental and company-wide change of existing practices aimed at improving company performance"*.

According to Chang (2005, p. 414), *"The continuous improvement cycle consists of establishing customer requirements, meeting the requirements, measuring success, and continuing to check customers' requirements to find areas in which improvements can be made"*. Chan's perspective of continuous improvement considered incremental client satisfaction. In exceeding clients' expectations, there needs to be a constant review of what went wrong and right and feedforward into future activities.

Bhuiyan, Baghel and Wilson (2006, p. 671) defined continuous improvement as a *"culture of sustained improvement aimed at eliminating waste in all organisational systems and processes and involving all organisational participants"*. Their inclusion of waste reduction in continuous improvement practices stems from lean thinking in organisations. Lean involves waste reduction and it is an integral part of continuous improvement.

Manos (2007, p. 47) definition of the concept as: *"subtle and gradual improvements that are made over time"* also considers change management in the process of continuous improvement and development. Although Manos's definition seems generic, there is a clear indication of incrementality in the process of attaining continuous improvement.

Overall, continuous improvement may seem to be a generic terminology. Yet, in its application, there are focal areas that articulate its innate meanings. Continuous improvement is the process of incrementally enhancing structures, services, products, quality, processes, and policies. For example, when continuous improvement is applied in an organisation, change management must be considered. In activities and processes, a systemic approach will be adopted. Continuous improvement is a philosophy that cyclically creates a steady non-ending change with the purpose of continually improving systems, methods, and processes.

1.1.2 Deming's cycle

Deming's cycle indicating the plan-do-check-act (PDCA) principle was designed by Dr Edward Deming as a tool for implementing continuous improvement within activities, as indicated in Figure 1.1. The first phase of continuous improvement must start with the identification of problems, waste, or non-value-adding activities. This will be followed by a plan to reduce the identified setbacks.

The planned solutions are tested in the "do" phase. A testing stage is required to fully understand the effect of the solution on a small-scale level before implementing it on a larger scale. Therefore, there is a need to check and analyse the outcomes of the small-scale solutions on the iden-tified challenges. The final phase involves taking a larger scale corrective action to resolve existing challenges. Further challenges are identified and the PDCA process continues in an iterative manner. The PDCA cycle can be used when embarking on new improvement projects; in develop-ing a new design, concept or process; setting out a new repetition work or process; planning data extraction and analysis in order to ascertain the causation of challenges; implementing change; and quantitatively working towards the elimination of waste.

It is essential to note that it is practically impossible to embed contin-uous improvement in a process without first applying it in the system. A new system in this instance can be a sector such as the construction industry, construction companies, sub-sectors such as manufacturing in the construction industry and processes such as design and cost planning.

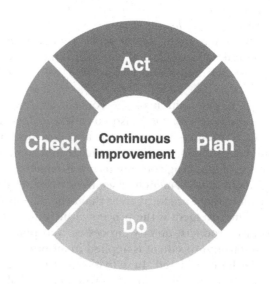

Figure 1.1 Deming's cycle.

Source: Authors.

In achieving continuous improvement, associated concepts will be elucidated and discussed in subsequent sections.

1.1.3 Nine principles of continuous improvement

For a practical implementation of continuous improvement in a system, some set of principles must be applied in an organisation. These principles range from organisational learning to productivity enhancement through waste elimination, process improvement, and prioritisation of activities.

1.1.3.1 Creating a learning organisation: Practical learning

Continuous improvement requires constant learning. The constancy of learning depends on the ambition, targets, and financial commitment an organisation is willing to commit towards training programmes. Many policy makers at the management level have used the term "knowledge-creating companies" or "knowledge-based organisations" to redefine the context of continuous improvement in their organisations. Irrespective of the nomenclature used, the process of creating a learning organisation involves knowledge capturing, knowledge management and dissemination, and shared interpretation. Many construction organisations have historical information in the form of drawings, reports, financial statements, and performance of previous projects, existing without an organised format of feedforwarding into future projects. Learning organisations exist for the purpose of creating new insights into existing processes and invariably influence behavioural change within the learning environment, in this case, the organisation. Continuous improvement can be described as having been successfully implemented in such organisations when existing knowledge and data are explored and utilised to create a competitive advantage. In an era in which construction businesses are increasingly competing for projects, training, research and innovation have become important factors in spurring organisational growth, thereby boosting competitiveness.

1.1.3.2 Waste reduction and elimination

There are two types of wastes on which continuous improvement must focus for the purpose of reduction and elimination. The first, physical waste, is tangible and is generated because of the production process, whereas the second type is non-physical waste which is intangible. Many academic studies have focussed on physical waste while neglecting the non-physical waste. Omotayo, Olanipekun, Obi and Boateng (2020b) identified overhead costs associated with planning, plant hire, labour payment, construction material deliveries, and payment of subcontractors and suppliers as belonging to the non-physical waste category in their study on the incremental reduction of non-physical waste in the construction industry. The hallmark of lean thinking is waste reduction and elimination. The concept of waste in continuous improvement relates to all forms of non-value-adding activities. Consequently, it is difficult to quantify non-physical waste unless it is attached to financial outcomes. For example, poor negotiation with a subcontractor can lead to excess charges and overall construction costs.

Material waste is easily quantifiable and may be reduced with an additional focus using any waste management plan or the PDCA cycle. In the United Kingdom (UK), 90 million tonnes of construction waste are generated annually. These wastes are generated from packaging, leftovers from construction materials, design error or changes, poor storage, pilfering, and handling of materials.

The type of activities contributing to material waste depends on the nature of the construction industry. In construction industries situated within developing country contexts such as that of Nigeria, the significant causes of construction waste on-site include poor allocation of resources; poor recording keeping; vandalism, variation, and rework; damage as a result of weather or mishandling; damage as a result of transportation; composite and design of the building; material supplied and used on-site and site office waste. The materials which generate waste on-site include concrete, wood, metals, plastic, tiles, insulations, paints, soil and stones, ceramics, glass, and bricks. Waste generation on construction sites should be avoided as the effect can lead to cost and time overruns. In some cases, it can abruptly end the project. Waste reduction and elimination through continuous improvement-based reduction processes provide enhanced profitability, sustainable construction, improved value, and client satisfaction. Nonetheless, these benefits depend on the efficacy of the continuous improvement concepts implemented. For instance, the identification of waste for continual reduction purposes using Kaizen costing depends on the supply chain, workers on-site, payment by the client, purchase orders, and actions are taken on site.

1.1.3.3 Employee engagement

The continuous improvement principles of learning organisations, waste reduction, and elimination are driven by employee engagement. Employee engagement has been described as a management approach to creating an enabling environment for the enshrinement of organisational core values and objectives, which invariably leads to improvement well-being (Brajer-Marczak, 2015). When employees have been motivated to the right levels within the work environment, continuous improvement will be easily facilitated. One of the key attributes of continuous improvement is looking for a new and better way of conducting repetitive tasks. These tasks include service or product development under different working conditions and environments. Employee engagement in attaining continuous improvement may be viewed from the perspective of problem-solving. When employees and employers constantly search for opportunities to eliminate waste and adopt new methods and techniques of improving their performance, continuous improvement will be realised over time. Becker, Prikladnicki, and Audy (2008) opined that minor improvements may pervade inter-departmental and interorganisational processes with micro-implementation within teams in an organisation. This approach will create minute changes in inter- and intra- organisational teams over a period of time, hence bringing about the emergence of a new organisational culture. The overall goal of employee engagement in continuous improvement is motivation, enhanced productivity, and cultural changes.

1.1.3.4 Increased productivity with less financial investment

There is a plethora of evidence to support increased productivity where continuous improvement has been implemented (Jin & Doolen, 2014; Suárez-Barraza, Ramis-Pujol & Llabrés, 2009). One of the main questions raised during the implementation of continuous improvement is the probable high cost of training and re-training programmes. Increased productivity with less financial investment in training programmes is a key concept of continuous improvement. The cost of continuous improvement training programmes is reduced when the culture of continuous improvement has permeated all managerial levels of an organisation. The cost of production can continually be reduced when non-value-adding activities are reduced. This creates less investment in the overall process of improvement and more profit.

Robert and Granja (2006) opined that the actual cost can be reduced with the same or reduced investment in the production of a product. Robert and Granja (2006) also explained that when production commences, the targeted profit margin is used as a benchmark. Kaizen costing, which is also continuous cost improvement, begins with unit 1 production. In the early stages of construction, namely units 1, 2, and 3, the idea will be to maintain the cost before reduction. Units n2, n3, and n are where continuous cost improvement is achieved through the elimination of non-value-adding activities and cost leakages on the site (Robert & Granja, 2006). The process continues until the end of the project. In continuous improvement, construction activities are viewed as a standardised production process, similar to what may be obtainable in a factory setting. In reality, construction activities have their own unique environment. The peculiar nature of the construction environment is a major barrier facing the complete adoption of continuous improvement on construction sites. Productivity enhancement through continuous improvement uses the same level of investment to create a high-profit margin.

1.1.3.5   Continuous improvement in any location

Continuous improvement can be practised in the office space or production line. The concept of *Gemba Kaizen* refers to continuous improvement in a working environment (Imai, 1992; Omotayo & Kulatunga, 2017). Continuous improvement in the place of work depends on an organisation's mission statement or core values, communication approach, organisational structure and complexity, waste reduction policy, knowledge of continuous improvement, and the process of feedforwarding outcomes of post-project reviews.

There are no limitations as to where and how continuous improvement can be applied. Private and public organisations have redefined their mission statement along the lines of waste reduction, employee engagement, and knowledge management. For instance, in the office environment, the concept of reducing paper trails for electronic copies is a simple practice which has changed the behaviour of employees and management staff. Furthermore, in the production line, the standardisation of repeated activities has proven to eliminate waste and enhance productivity. Continuous improvement is still an emerging concept in many developing economies and small and medium-scale companies are also increasingly adopting this new concept (Imai, 1992; Mĺkva, Prajová, Yakimovich, Korshunov & Tyurin, 2016). In China, India, Brazil, Nigeria, South Africa, the UK, some European countries, Canada, and the USA, continuous improvement has become a mantra for development and innovation in management and process improvement.

1.1.3.6   Disclosure of improvement through transparent communication

One of the tools facilitating continuous improvement is transparent communication. Communication can only be effective when there is a level of standardisation of processes and roles. In the construction sector, construction activities have been described as a manufacturing process by Koskela *et al.* (2002). Standardisation in construction is evident in modular and volumetric construction. Processes that demand repetition in the form of drawings, planning, and reviews may also benefit from standardisation. Effective communication is essential for continuous improvement (Shang & Pheng, 2014). The communication approach may be top-down or bottom-up, specific, or non-specific. Irrespective of the adopted communication approach, there should be a transparent disclosure of the positive and negative outcomes of continuous improvement. The essence of transparent disclosure in continuous improvement is motivation. Motivation is an integral part of achieving continual improvement in an organisation. When the workforce receives transparent feedback about the outcomes of continuous improvement practices they were part of or knew about, they may be motivated to engage in continuous improvement processes established by the PDCA cycle.

1.1.3.7 Focus on areas of greatest need

As part of the PDCA cycle, the first step, which is the "plan", focusses on identifying problems and areas where improvements are required. The aim of continuous improvement is to create small improvements with minimal expenses. Hence, an assessment of an existential problem in a working environment, project, or process is required to ascertain its level of severity. For example, a construction project may prioritise the movement of construction materials around the site, time management of activities, defect reduction on construction sites, over-processing, and variation management. The focal areas of continuous improvement dictate the direction of improvement. In prioritising the areas of greatest needs for improvement, benchmarking tools, risk management, progress reports, and feedback loops are significant in decision-making.

1.1.3.8 Focus on process improvement

Continuous improvement should always focus on processes for practical implementation. The planning and execution process can be enhanced when there is a cultural change towards continually improving processes that dictate the direction of outputs. Good timekeeping can lead to additional productive time. Thorough and in-depth analysis before decision-making may also eliminate waste. Since the overall goal of continuous improvement is waste reduction and elimination, well-informed decision-making can improve processes. In construction projects off-site, volumetric processes and the application of new technology have become process improvement mechanisms (Henderson & Ruikar, 2010; Wong, 2007; Yisa, Ndekugri & Ambrose, 1996). Process improvement in the construction industry is linked with new construction technology. In the later chapters of this book, the nature and impact of new construction processes will be associated with continuous improvement.

1.1.3.9 Prioritising employee improvement

The final feature of continuous improvement is employee improvement. When employees are engaged in process improvement, training programmes, waste reduction, and elimination, the resultant effect will be employee improvement. The main drivers of learning organisations are the employees. Employees' improvement usually leads to improved client satisfaction, production quality, services, and productivity. The areas of greatest need in an organisation always involve the employees. An employee performance improvement plan (PIP) is a structured document which can be used to continually monitor behavioural and achievement goals and talent development. Intermittent training can take place with a capability maturity model management approach towards employee development.

Capability maturity models can aid employee and organisational improvement when there is a similar framework associated with the existing core values. Additionally, a value stream map for employee development has been proven to enhance quality, products, and services (Chen, Li & Shady, 2010). Value stream mapping does not only reduce production lead times but also increases the productivity of employees.

1.1.4 Creating a value stream mapping

Value stream comprises all information and materials required for the manufacturing process of a product. Value stream mapping is, therefore, the transfer of information into a framework, in this case, a map that provides current and future details in order to produce a manufacturing system (Riezebos & Huisman, 2020). Value stream mapping is a lean manufacturing concept that is based on a standardisation of operation routine. Standardisation is required for the continuous improvement of processes, while in value stream mapping, information required for the development of routine sheets must include the time it takes a worker to complete a manufacturing activity, machine operating times, and manual processing time. Value stream mapping is a continuous improvement tool that the construction sector may adopt on construction sites for downtime reduction, process improvement, and waste reduction (Elizar, Suripin & Wibowo, 2017).

1.1.5 The merits and demerits of continuous improvement

Continuous improvement has proven to aid the effective utilisation of available resources in an organisation. The process involves the elimination of non-productive activities, which can increase during working hours. Employee productivity has been improved using continuous improvement techniques. Utari (2011) studied the use of continuous improvement in PT Coca-Cola Bottling Indonesia Central, Sumatera. Findings from that study highlighted the process of eliminating rejected products which gave rise to significantly increased profit margins in the company. This process consists of value stream mapping and an identification of what the consumers want while disregarding any component or product which does not add value. Ellram (2006) and Singh and Singh (2012) identified the benefits of continuous improvement as being the following:

i   Continuous incremental improvement of the product cost.
ii Incremental reduction of waste during the construction stage.
iii Cost reduction in the supply chain.
iv Profit improvement.
v   Improvement of competitive advantage of a small- and medium-scale companies.

vi Integrated with organisations, values, mission, and developmental objectives.

vii Improvement of budgeting systems and reduction in non-productive working hours.

A significant benefit of continuous improvement in organisations is the low level of financial investment required to integrate its concepts (Brajer-Marczak, 2015). Continuous improvement also affords employees the opportunity to make their ideas heard by the management, thereby creating a healthy workplace. Organisations that fail to incorporate continuous improvement have the traits of re-strategising, re-organising, re-structuring, and impulsivity towards negative external and internal changes. From a micro-perspective, continuous improvement brings out the best in talents within an organisation. Continuous improvement practices in the construction industry remain sparse when compared with other sectors. However, evidence abounds in lean construction practice case studies in the industry.

1.1.6 Evidence of continuous improvement in the construction industry

Continuous improvement approaches are still emerging in the construction industry.

Research into the application of continuous improvement in the construction industry had suggested its merger with construction business operations, construction management function, construction business ethics, and construction cost management (Omotayo, Kulatunga & Bjeirmi, 2018). Additionally, continuous improvement has been combined with target costing in lean construction activities for cost reduction and maintenance (Robert & Granja, 2006). The main branches of application have been divided into the planning or organisational management and execution or construction phases. Table 1.1 summarises the applications of continuous improvement in the construction industry as used in various countries.

Table 1.1 displays a summary of studies in the application of continuous improvement in the construction industry. These studies prioritised the creation of theoretical frameworks and the exhibition of practical application case studies. Based on the contents of the studies presented in Table 1.1, the application of continuous improvement in the construction industry can be categorised into learning organisations, process improvement, quality management, physical waste reduction, non-physical waste reduction, productivity enhancement, construction project management improvement, building information model (BIM), digital design improvement, customer behaviour improvement, and construction cost management. These applications are mainly in the areas of building construction;

Table 1.1 Studies into the application of continuous improvement in the construction industry

Application in construction	Construction sector	Construction theoretical framework or direct application	Country	Source
Construction learning organisations	Road infrastructure	Theoretical framework	Netherlands	Gieskes and Ten Broeke (2000)
Process improvement	Building	Theoretical framework	Chile	Serpell et al. (1996)
Process improvement and organisational culture	Construction organisations	Theoretical framework	UK	Jeong et al. (2004)
Cost management enhancement	Building	Theoretical framework	Nigeria	Omotayo et al. (2020a)
Cost management enhancement	Building	Theoretical framework	Chile	Robert and Granja (2006)
Physical waste reduction in construction processes	Building	Theoretical framework	Nigeria	Omotayo et al. (2019)
Process improvement and waste reduction	Building	Theoretical framework and practical application	Brazil	Vivan et al. (2015)
Non-physical waste reduction	Building	Theoretical framework	Nigeria	Omotayo et al. (2020b)
Quality management and organisational growth	General construction	Theoretical framework	Nigeria	Jimoh et al. (2019)
On-site labour productivity improvements	Power, petroleum, and petrochemical projects	Practical application	USA	Gouett et al. (2011)
Quality management	Building	Theoretical framework	Hong Kong	Tam et al. (2000)
Building information modelling (BIM)	General construction	Theoretical framework	UK	Mubarak and Mustafa (2015)
Construction management and productivity of bricklayers	Building	Practical application	Brazil and UK	Aguinaldo et al. (2000)
Off-site construction	Building	Practical application	Sweden	Meiling et al. (2012)
Productivity and quality management	Building	Practical application	Chile	De Solminihac et al. (1997)
Digital design and BIM improvement	Building	Theoretical framework and Practical application	UK	Tetik et al. (2019)
Construction project management improvement	Building	Theoretical framework	Finland	Savolainen et al. (2015)
Organisational management	Concrete production	Practical application	Finland	Savolainen (1999)
Customer behaviour and total quality management	General construction	Theoretical framework	Iran	Javad et al. (2019)

infrastructure; power, petroleum, and petrochemical projects; concrete production; and general construction. The categorisation of general construction was derived from a continuous improvement framework designed for non-specific construction activities. The aforementioned continuous improvement applications in the construction industry are described below.

1.1.6.1 Learning organisations in the construction industry

Gieskes and Ten Broeke (2000) and Jeong, Siriwardena, Amaratunga, Haigh and Kagioglou (2004) all documented theoretical frameworks for the improvement of the management function and construction organisations in the Netherlands and UK, respectively. The main feature of continuous improvement is the creation of learning organisations. Construction businesses have an opportunity to harness their vast repository of data and information. Construction learning organisations serve as the bedrock for the deployment of any other form of continuous improvement application.

1.1.6.2 Process improvement

Construction learning organisations have improvements in their administration, management, and production processes. Serpell, Alarcón and Ghio (1996) and Jeong *et al.* (2004) developed theoretical frameworks for construction process improvement. Vivan, Ortiz and Paliari (2015) provided a theoretical framework and practical application of continuous improvement in Brazil. Vivan *et al.* (2015) further identified the benefits of continuous improvement in building construction as including the reduction of downtimes, early delivery of materials from the suppliers, and efficiencies in workflow. Process improvement is a benefit and serves as evidence of the utility of continuous improvement. The dynamics of construction processes can be eased when there are repeatable processes that can be performed in a more efficient manner. The concept of process improvement in the construction industry starts with the identification of repeatable activities during the planning, tendering, and construction phases. Physical and non-physical activities in these activities are then identified for elimination. Furthermore, a new efficient approach is designed and implemented. These processes are then repeated.

1.1.6.3 Physical waste reduction

Physical waste is tangible, solid material waste which may be plastic, paper, concrete, and every other construction material which is not used. Although the physical waste reduction in the construction industry can be mitigated through the application of circular construction methods and practices, continuous improvement as a lean concept provides a direct and practical strategy towards its reduction and elimination. Omotayo, Boateng, Osobajo, Oke and Obi (2019) developed a theoretical framework for the implementation of continuous improvement in the Nigerian construction industry. This framework further introduced the capability maturity framework for construction businesses in Nigeria. The capability maturity framework developed in that study elucidated the guidelines for construction companies to attain optimal levels of continuous improvement implementation over a particular time interval.

1.1.6.4 Non-physical waste reduction

Non-physical wastes in construction are activities that do not add value to the overall construction process. Examples of these may be delays in delivery or preparing tender documents, poor project planning, rework, and correction of errors. Omotayo *et al.* (2020b) suggested a systems thinking attitude towards the mitigation of non-physical waste in construction as a backdrop for enhancing physical waste management on sites. Omotayo *et al.* (2020b) provided a theoretical framework for managing the incidence of non-physical waste, such as high overhead costs, variations, drawing reviews, preliminary items of work, supply, and subcontractors' cost. Simple activities making up intangible waste are mitigated, while the organisational culture can potentially gravitate towards the reduction of physical waste.

1.1.6.5 Improving quality management

Total quality management in the construction industry has been identified as a key benefit and an outcome of continuous improvement by Jimoh, Oyewobi, Isa and Waziri (2019), Tam, Deng, Zeng and Ho (2000), and Javad, Nafiseh and Paria (2019) in Nigeria, Hong Kong, and Iran, respectively. These authors noted that the quality of the construction production process, product and customer relationship can be improved through the introduction of a framework to guide construction project managers in executing their projects. In Hong Kong, Tam *et al.* (2000) discussed the emergence of new quality standards in the country and how a continuous improvement framework can augment existing standards through productivity, technology, and continuous learning.

1.1.6.6 Productivity enhancement

Gouett, Goodrum and Caldes (2011) documented a practical application of how continuous improvement has helped to enhance on-site labour productivity of power, petroleum, and petrochemical projects in the US. The case studies conducted by Gouett *et al.* (2011) revealed higher productivity outcomes in construction projects where the labourers were instructed along continuous improvement practices such as the PDCA cycle and waste minimisation. De Solminihac, Bascunan and Edwards (1997) also provided a practical example of how continuous improvement enhanced the construction of buildings in Chile. Practical applications which culminated in increased productivity of labourers and construction project management in the US and Chile were documented in these studies. There is a possibility to transfer these applications into other construction industry contexts through a tailored framework.

1.1.6.7 *Construction project management improvement*

Savolainen (1999) and Savolainen, Kähkönen, Niemi, Poutana and Varis (2015) studied the application of continuous improvement processes in concrete production and building projects in Finland. Findings from both studies created a theoretical and practical framework for organisational and construction project management improvement. Aguinaldo, Andrew and Carlos (2000) also recognised the practical application of continuous improvement of construction project management in Brazil and the UK. They studied how managers were able to adopt the concepts of continuous improvement in delivering a building project. The outcome of that study ensured innovation, better workflow, time management, and cohesiveness among construction managers. The enhancement of construction project activities is possible with the inclusion of lean principles and the standardisation of activities.

1.1.6.8 *Building information model (BIM) and digital design improvement*

The impact of the emergence of digital technologies and building information modelling in design, management, and construction has been acknowledged in the construction industry. Continuous improvement offers an opportunity to adopt BIM and other digital technologies easily. For instance, Tetik, Peltokorpi, Seppanen and Holmstrom (2019) articulated practical case studies on how continuous improvement facilitated the use of BIM and digital technologies throughout the project life cycle in the UK. Mubarak and Mustafa (2015) also produced a theoretical framework for the application of continuous improvement in any construction project. The works of Tetik *et al.* (2019) and Mubarak and Mustafa (2015) are indications of the relevance of continuous improvement in the construction industry. BIM maturity models are inclusive of sustainable and innovative practices which encourage continuous improvement. Innovation in the Internet-of-things (IoT) and associated digital technologies such as mobile project management in construction are evidence of how continuous improvement exists in information and communication technology and can also be transferred into the construction industry.

1.1.6.9 Customer behaviour improvement

Quality management and customer behaviour were the focus of the study by Javad *et al.* (2019) into continuous improvement practice in Iranian construction organisations. The authors suggested that the deployment of a theoretical framework for changing organisational behaviour from tasks to processes can have a direct influence on the quality of products and services delivered by any organisation.

1.1.6.10 Off-site construction

The earliest applications of continuous improvement began with the off-site manufacturing sector. Meiling, Backlund and Johnsson (2012) documented a practical application of continuous improvement in the off-site construction of buildings in Sweden. These scholars opined that continuous improvement in the construction sector must incrementally influence processes, people, and long-term thinking. Off-site construction and the enhancement of management principles served as an enabler for every other concept of continuous improvements, such as physical waste minimisation, timely delivery of products, quality management, and cost reduction or maintenance.

1.1.6.11 Construction cost management

Through the aid of systems thinking, Omotayo *et al.* (2020a) developed a theoretical framework for the integration of continuous improvement with cost management in the Nigerian construction industry. Construction costs are aggravated by improper management of overhead cost, variations as well as the cost associated with preliminary items of works, such as electricity, insurance, security, and water. Omotayo *et al.* (2020a) opined that an incremental minimisation of construction costs from the government, construction professional regulatory bodies, construction project organisations, and managers on-site will lead to the actualization of a continuous cost improvement regime. The aforementioned stakeholders are crucial to creating a new culture of improvement in organisations before the production phase. Robert and Granja (2006) illustrated how target costing and continuous cost improvement in construction should be introduced from the planning phases of construction for the reduction and maintenance of construction cost. The targeted cost is continually reduced by ensuring lower cost targets, thereby increasing the contractor's profit margin.

1.1.7 Continuous improvement and construction cost management

There is a general assumption of continuous improvement in construction cost management leading to reduced costs at the expense of quality.

However, studies conducted by Robert and Granja (2006), Vivan *et al.* (2015), and Omotayo *et al.* (2020a) all highlighted improved levels of quality and improved employee–employer relationships, client satisfaction, and organisational growth as key benefits of continuous improvement. The following steps can be adopted for a clear integration of continuous improvement in construction cost management:

 i   Defining the project objectives clearly.
 ii Standardising the construction process.
 iii Identifying the cost alternative.
 iv Associating cost targets with construction activities.
 v   Developing a PDCA cycle for cost management activities using the bills of quantities or any related cost plan.
 vi Assigning a continuous cost improvement champion.
 vii Creating a continuous cost improvement plan or framework.
 viii Executing the framework during construction activities for cost minimisation.
 ix Monitoring and reviewing the outcome of the continuous improvement plan through weekly or monthly site meetings.
 x   Conducting a post-project review with an emphasis on the construction cost at the end of the project.

References

Aguinaldo, S., Andrew, P. J. and Carlos, T. F. (2000). Setting stretch targets for driving continuous improvement in construction: Analysis of Brazilian and UK practices. *Work Study*, 49(2), pp. 50–58. doi: 10.1108/00438020010311179

Becker, A. L., Prikladnicki, R., & Audy, J. L. N. (2008). Strategic alignment of software process improvement programs using QFD. In *Proceedings of the 1st International Workshop on Business Impact of Process Improvements* (pp. 9–14). https://doi.org/10.1145/1370837.1370840

Bessant, J., Caffyn, S., Gilbert, J., Harding, R. and Webb, S. (1994). Rediscovering continuous improvement. *Technovation*, 14(1), pp. 17–29. doi: 10.1016/0166-4972(94)90067-1

Bessant, J., Caffyn, S. and Gallagher, M. (2001). An evolutionary model of continuous improvement behaviour. *Technovation*, 21(2), pp. 67–77. doi: http://dx.doi.org/10.1016/S0166-4972(00)00023-7

Bhuiyan, N., Baghel, A. and Wilson, J. (2006). A sustainable continuous improvement methodology at an aerospace company. *International Journal of Productivity and Performance Management*, 55(8), pp. 671–687. doi: 10.1108/17410400610710206

Boer, H. and Gertsen, F. (2003). From continuous improvement to continuous innovation: A (retro)(per)spective. *International Journal of Technology Management*, 26(8), pp. 805–827.

Brajer-Marczak, R. (2015). Employee engagement in continuous improvement of processes. *Management*, 18(2), pp. 88–103. doi: 10.2478/manment-2014-0044

Brunet, A. P. and New, S. (2003). Kaizen in Japan: An empirical study. *International Journal of Operations & Production Management*, 23(12), pp. 1426–1446.

Chang, H. H. (2005). The influence of continuous improvement and performance factors in total quality organization. *Total Quality Management & Business Excellence*, 16(3), pp. 413–437.

Chen, J. C., Li, Y. and Shady, B. D. (2010). From value stream mapping toward a lean/sigma continuous improvement process: An industrial case study. *International Journal of Production Research*, 48(4), pp. 1069–1086. doi: 10.1080/00207540802484911

Dahlgaard, J., Kristensen, K. and Kanji, G. (2010). Total quality management and education. *Total Quality Management*, 6, pp. 445–456. doi: 10.1080/09544129550035116

Deming, W. E. (1982). *Out of the crisis*. Cambridge, MA: Massachusetts Institute of Technology Center for Advanced Engineering Study. https://mitpress.mit.edu/books/out-crisis

De Solminihac, T. H., Bascunan, R. and Edwards, L. G. (1997). *Continuous improvement in construction management and technologies: A practical case*. Santiago, Chile, and Balkema, Rotterdam, The Netherlands: Luis Alarcon School of Engineering, Catholic Univ. of Chile, pp. 240–262.

Elizar, E., Suripin, S. and Wibowo, M. A. (2017). The concept of value stream mapping to reduce work-time waste as applied in the smart construction management. In *Proceedings of the 3rd International Conference on Construction and Building Engineering (Conbuild) 2017. AIP Conference Proceedings*, 1903(1), p. 070010. doi: 10.1063/1.5011579

Ellram, L. M. (2006). The implementation of target costing in the United States: Theory versus practice. *The Journal of Supply Chain Management*, 42(1), pp. 13–26.

Gieskes, J. and Ten Broeke, A. (2000). Infrastructure under construction: Continuous improvement and learning in projects. *Integrated Manufacturing Systems*, 11(3), pp. 188–198. doi: 10.1108/09576060010320425

Gouett, M. C., Goodrum, P. M. and Caldes, C. H. (2011). Activity analysis for direct-work rate improvement in construction. *Journal of Construction Engineering and Management*, 137(12), pp. 1117–1124. doi: 10.1061/(asce)co.1943-7862.0000375

Henderson, J. R. and Ruikar, K. (2010). Technology implementation strategies for construction organisations. *Engineering, Construction and Architectural Management*, 17(3), pp. 309–327. doi: 10.1108/09699981011038097

Imai, M. (1992). Comment: Solving quality problems using common sense. *International Journal of Quality & Reliability Management*, 9(5). doi: 10.1108/EUM0000000001655

Javad, E. M., Nafiseh, N. and Paria, S. (2019). Measuring the impact of soft and hard total quality management factors on customer behavior based on the role of innovation and continuous improvement. *The TQM Journal*, 31(6), pp. 1093–1115. doi: 10.1108/TQM-11-2018-0182

Jeong, K. S., Siriwardena, M. L., Amaratunga, R. D. G., Haigh, R. P. and Kagioglou, M. (2004). Structured process improvement for construction enterprises (SPICE) level 3: Establishing a management infrastructure to facilitate process improvement at an organisational level. In *Proceedings of 1st International SCRI Research Symposium*, 30–31 March, Manchester, UK.

Jimoh, R., Oyewobi, L., Isa, R. and Waziri, I. (2019). Total quality management practices and organizational performance: The mediating roles of strategies for continuous improvement. *International Journal of Construction Management*, 19(2), pp. 162–177. doi: 10.1080/15623599.2017.1411456

Jin, H. W. and Doolen, T. L. (2014). A comparison of Korean and US continuous improvement projects. *International Journal of Productivity and Performance Management*, 63(4), pp. 384–405. doi: 10.1108/IJPPM-01-2013-0012

Juergensen, T. (2000). *Continuous improvement: Mindsets, capability, process, tools and results*. Indianapolis: The Juergensen Consulting Group Inc.

Kapur, A., Adair, N., O'Brien, M., Naparstek, N., Cangelosi, T., Zuvic, P., Sherin, J., Meier, S., Meier, J. Bloom, B. and Potters, L. (2016). Improving efficiency and safety in external beam radiation therapy treatment delivery using a Kaizen approach. *International Journal of Radiation Oncology – Biology – Physics*, 96(2, Supplement), p. S73. doi: 10.1016/j.ijrobp.2016.06.186

Koskela, L., Ballard, G., Howell, G. and Tommelein, I. (2002). The foundation of lean construction. In R. Best and G. de Valence (Eds.). *Design and construction: Building in value*. Oxford: Butterworth Heinemann, pp. 211–226.

Maarof, M. G. and Mahmud, F. (2016). A review of contributing factors and challenges in implementing Kaizen in small and medium enterprises. *Procedia Economics and Finance*, 35, pp. 522–531. http://dx.doi.org/10.1016/S2212-5671(16)00065-4

Macomber, H., Howell, G. and Barberio, J. (2007). Target-value design: Nine foundational practices for delivering surprising client value. Lean Project Consulting, Vol. 80027, pp. 1–2, Louisville.

Manos, A. (2007). Lean lessons: The benefits of Kaizen and Kaizen events. *Quality Progress*, 40(2), p. 47.

Meiling, J., Backlund, F. and Johnsson, H. (2012). Managing for continuous improvement in off-site construction: Evaluation of lean management principles. *Engineering, Construction and Architectural Management*, 19(2), pp. 141–158. doi: 10.1108/09699981211206089

Mĺkva, M., Prajová, V., Yakimovich, B., Korshunov, A. and Tyurin, I. (2016). Standardization – One of the tools of continuous improvement. *Procedia Engineering*, 149, pp. 329–332. http://dx.doi.org/10.1016/j.proeng.2016.06.674

Mubarak, A. A. and Mustafa, A. (2015). BIM for client organisations: A continuous improvement approach. *Construction Innovation*, 15(4), pp. 402–408. doi: 10.1108/CI-04-2015-0023.

Narasimhan, K. (2002). Implementing the capability maturity model. *The TQM Magazine*, 14(2), pp. 133–135. doi: 10.1108/tqmm.2002.14.2.133.1

Omotayo, T., Awuzie, B., Egbelakin, T., Obi, L. and Ogunnusi, M. (2020a). AHP-systems thinking analyses for Kaizen costing. *Buildings*, 10(12), p. 230.

Omotayo, T. S., Boateng, P., Osobajo, O., Oke, A. and Obi, L.I. (2019). Systems thinking and CMM for continuous improvement in the construction industry. *International Journal of Productivity and Performance Management*, 69(2), pp. 271–296. doi: 10.1108/IJPPM-11-2018-0417

Omotayo, T. and Kulatunga, U. (2017). A continuous improvement framework using IDEF0 for post-contract cost control. *Journal of Construction Project Management and Innovation*. University of Johannesburg, 7(1), pp. 1807–1823. https://africaneditors.org/journal/JCPMI/abstract/76552-223700

Omotayo, T. S., Kulatunga, U. and Bjeirmi, B. (2018). Critical success factors for Kaizen implementation in the Nigerian construction industry. *International Journal of Productivity and Performance Management*, 67(9), pp. 1816–1836. doi: 10.1108/IJPPM-11-2017-0296

Omotayo, T., Olanipekun, A., Obi, L. and Boateng, P. (2020b). A systems thinking approach for incremental reduction of non-physical waste. *Built Environment Project and Asset Management*, 10(4), pp. 509–528. doi: 10.1108/BEPAM-10-2019-0100

Prashar, A. (2014). Process improvement in farm equipment sector (FES): A case on Six Sigma adoption. *International Journal of Lean Six Sigma*, 5(1), pp. 62–88. doi: 10.1108/IJLSS-08-2013-0049

Riezebos, J. and Huisman, B. (2020). Value stream mapping in education: Addressing work stress. *International Journal of Quality & Reliability Management*, 38(4). doi: 10.1108/IJQRM-05-2019-0145

Robert, G. T. and Granja, A. D. (2006). Target and Kaizen costing implementation in construction, In *Proceedings of the 14th Annual Conference on Lean Construction*, Santiago.

Sanchez, L. and Blanco, B. (2014). Three decades of continuous improvement. *Total Quality Management & Business Excellence*. Routledge, 25(9–10), pp. 986–1001. doi: 10.1080/14783363.2013.856547

Savolainen, T. I. (1999). Cycles of continuous improvement: Realizing competitive advantages through quality. *International Journal of Operations & Production Management*, 19(11), pp. 1203–1222. doi: 10.1108/01443579910291096

Savolainen, J., Kähkönen, K., Niemi, O., Poutana, J. and Varis, E. (2015). Stirring the construction project management with co-creation and continuous improvement. *Procedia Economics and Finance*, 21, pp. 64–71. https://doi.org/10.1016/S2212-5671(15)00151-3

Serpell, A., Alarcón, L. and Ghio, V. (1996). A general framework for improvement of the construction process. In Annual Conference of the International Group for Lean Construction, 4, Birmingham, UK.

Shang, G. and Sui Pheng, L. (2014). Barriers to lean implementation in the construction industry in China. *Journal of Technology Management in China*, 9(2), pp. 155–173. doi: 10.1108/JTMC-12-2013-0043

Singh, J. and Singh, H. (2012). Continuous improvement approach: State-of-art review and future implications. *International Journal of Lean Six Sigma*, 3(2), pp. 88–111. doi: 10.1108/20401461211243694

Singh, J. and Singh, H. (2015). Continuous improvement philosophy – literature review and directions. *Benchmarking: An International Journal*, 22(1), pp. 75–119. doi: 10.1108/BIJ-06-2012-0038

Suárez-Barraza, M. F. and Ramis-Pujol, J. (2010). Implementation of lean-Kaizen in the human resource service process: A case study in a Mexican public service organisation'. *Journal of Manufacturing Technology Management*, 21(3), pp. 388–410. doi: 10.1108/17410381011024359

Suárez-Barraza, M. F., Ramis-Pujol, J. and Kerbache, L. (2011). Thoughts on Kaizen and its evolution: Three different perspectives and guiding principles. *International Journal of Lean Six Sigma*, 2(4), pp. 288–308. doi: 10.1108/20401461111189407

Suárez-Barraza, M. F., Ramis-Pujol, J. and Llabrés, X. T. (2009). Continuous process improvement in Spanish local government: Conclusions and recommendations. *International Journal of Quality and Service Sciences*, 1(1), pp. 96–112. doi: 10.1108/17566690910945895

Tam, C. M., Deng, Z. M., Zeng, S. X. and Ho, C. S. (2000). Quest for continuous quality improvement for public housing construction in Hong Kong. *Construction Management and Economics*, 18(4), pp. 437–446. doi: 10.1080/01446190050024851

Tetik, M., Peltokorpi, A., Seppanen, O. and Holmstrom, J. (2019). Direct digital construction: Technology-based operations management practice for continuous improvement of construction industry performance. *Automation in Construction,* 107, p. 102910. https://doi.org/10.1016/j.autcon.2019.102910

Utari, W. (2011). Application of kaizen costing as a tool of efficiency in cost of production at Coca Cola Bottling Indonesia. Central Sumatra, Indonesia Andalas University Padang.

Vivan, A. L and Ortiz, F. A, Paliari, J. (2015). Model for Kaizen project development for the construction industry. *Gestão & Produção,* 23(2), pp. 333–349. doi.org/10.1590/0104-530X2102-15

Wong, C. H. (2007). ICT implementation and evolution: Case studies of intranets and extranets in UK construction enterprises. *Construction Innovation,* 7(3), pp. 254–273. doi: 10.1108/14714170710754740

Yisa, S. B., Ndekugri, I. and Ambrose, B. (1996). A review of changes in the UK construction industry: Their implications for the marketing of construction services. *European Journal of Marketing,* 30(3), pp. 47–64. doi: 10.1108/03090569610107427

2 Construction Cost Management Systems, Methods, and Techniques

Temitope Omotayo, Udayangani Kulatunga, and Bankole Awuzie

PART I

2.1 Introduction: Construction cost management

According to the Project Management Body of Knowledge (PMBOK), construction cost management can be described as a system of developing construction project estimates, budgets, and control of expenditure during construction (Owens, Burke, Krynovich and Mance, 2007). Clients' cost requirements must be articulated to ensure early cost advice and cost plan development. Construction cost ensures that clients' cost requirements are infused into the design, procurement selection, construction activities, and operation of the property. Therefore, the role of the quantity surveyor in construction cost management starts during the briefing stages of a project life cycle and extends to the in-use phase of a project where operation and maintenance are sacrosanct. However, researchers such as Rad (2002) and Venkataraman and Pinto (2011) opined that the purpose of construction cost management is geared towards monitoring on-site activities against planned budget for effective and efficient delivery of projects. Construction cost management extends beyond on-site cost monitoring of construction materials, labour, plant, equipment, and overheads as life cycle costing evaluation before and during construction has become an integral aspect of the process.

Construction cost management is an integrated information charter for managing financial dealings and processes in projects (Kern and Formoso, 2004). Its overall aim is to ensure that projects are delivered within a budgeted limit. Construction cost management depends on the managerial roles of a construction company; thus, the expertise of the cost consultant in a construction project will go a long way in determining the cost performance of a project. The decision-making processes in construction cost management depend on pro-activeness and the external factors influencing the construction business and projects. The nature of a procurement strategy also influences the system of cost management adopted for a construction project. For instance, the traditional procurement route will adopt the traditional cost management system where a bill of quantities document will

DOI: 10.1201/9781003176077-3

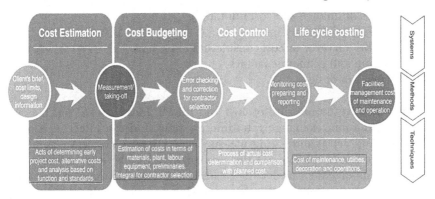

Figure 2.1 Construction cost management process.

Source: Authors.

be produced. A design and build procurement route will necessitate the early appointment of a contractor and a bill of approximate quantity will be used instead of a bill of quantities. Hence, construction cost management processes vary and depend on the nature of the procurement route, project, client, cost consultants, design team, and relevant stakeholders.

Figure 2.1 delineates construction cost estimation from budgeting, control, and life cycle costing. Cost management documents are prepared at every phase of the construction process. In the initial estimation phase, early cost advice is provided to the client by the quantity surveyor. Although the contractor produces his/her own estimates before and during construction, the working budget, which may be in the form of a bill of quantities, is the main tender cost document. Cost control activities are conducted to monitor project expenses and compare the actual cost with budgeted cost. The life cycle costing stage is also part of cost planning and construction, even though it is applied during the operational phase of a property.

The process of construction cost management involves estimating, budgeting, and controlling project cost (Jainendrakumar, 2015). The basic agenda of every construction cost management process is to prevent incidents of cost overrun. Construction cost management commences early during construction, as indicated in Figure 2.1. Tichacek (2006) noted that construction cost management begins with identifying important cost-generating activities during the execution phase of a project. The consciousness of cost generating activities during construction is aligned with construction material, labour, and plant costs. The main constraints faced by the quantity surveyor at the initial stages of the construction cost management process include the following:

i Lack of clear project scope during the briefing stage
ii Incomplete specifications

iii Sources of finance and cost limitations
iv Design requirements
 v Quantity surveyor or cost consultant's experience
vi Inflation and economic factors
vii Mistakes in design

The effective monitoring of cost and mindfulness of external factors such as inflation, exchange rate, interest rates and the effect of supply and demand act as mitigating measures for effective cost management. It is important to understand the processes and tools used in construction cost management under the headings of construction cost management systems, methods, and techniques. However, before this is done, the next section provides an insight into the difference between the concepts of construction cost, price, and value. Such a delineation is essential as it forms the underlying formwork upon which the concept of construction cost management is predicated.

2.1.1 Construction cost estimation

Cost planning activities lead to estimates for the project. The term "estimate" is used to describe cost planning outputs because the final documents are not going to bear the exact final construction cost. Consequently, the cost estimate is the amount of money required to complete a construction project (Oyedele, 2015). Cost estimates encompass planning the cost in terms of approximate estimation, taking-off from drawings, rates build up, and cost benchmarking activities. The process involves the incorporation of cost data for a careful allocation of project resources with their relevant activities (Kern and Formoso, 2004; Owens *et al.*, 2007). Budget estimates in the bills of quantities and approximate bills of quantities are well-established for informing the contractor, construction project manager and every other relevant stakeholder of where the costs should be directed to on the construction site. Firm price estimates are useful for the purpose of tendering and the not-to-exceed estimates are based on firm price estimates. Estimates which are categorised as firm price indicate an upper and lower cost limit that must not be exceeded by the contractor.

Ashworth and Perera (2015) noted that cost estimates can be classified as the following:

  I **Initial estimates:** In this type of estimate, preliminary feasibility studies and consultations to determine the cost limits of a project are conducted.
 II **Firm estimates:** This determines the cost information in brief for feasibility and viability appraisal.
III **Preliminary cost plan:** This cost plan is determined using previous or historical projects as a yardstick for creating an initial cost plan. This

cost plan is not the final cost plan and only provides an indication of cost limits. Improvements on this cost plan will be made once there is a detailed drawing.

IV **Final cost plan:** After having developed a preliminary cost plan, iterations on existing cost plans are made before the final cost plan. The final cost plan is produced once there are more details on a drawing and the rates in terms of cost per square meter are used against accurate quantities.

V **Cost checking stage:** The final costs are drawn from quantities taken from a drawing using the taking-off process. This process leads to final budget preparation after the rates are multiplied by measured quantities to produce the final construction cost.

Cost estimates are supposed to lead to the production of the working budget, in this case, the bills of quantities. The overall aim of the bills of quantities, which may be unpriced, serves to act as a tender document for contractor selection. Construction cost estimates which lead to the production of the working budget are not based on personal heuristics or historical data but on existing data extracted from detailed working drawings, construction price books and specifications. In instances where there are no detailed working drawings, the bill of approximate quantities may be produced from conceptual designs.

2.2 Differences between construction cost, price, and value

For an in-depth understanding of construction cost management terminologies and applications in the construction industry, it is necessary to establish a dissociation of basic terminologies such as price, cost, and value. In many research articles, books and the use of words, price and cost have been used interchangeably to provide the same meaning.

2.2.1 Cost

The Cambridge Dictionary (2021) defined cost as the amount of money required to be paid for a particular job, service, or production. Cost is associated with the act of producing or providing an output. Cost in the construction industry deals with the amount of money to be given up in order for a service or product to be rendered or delivered. Service and products in this regard may be tender documents and a building, respectively.

From a practical perspective, the cost is budgeted for using the bill of quantities, bills of approximate quantities, or elemental cost plans. In the process of construction, the cost will be invoiced as purchases of materials, labour payment, plant hire, and every other amount related to the production of a building, road, or bridge. At the point of purchase of building materials, it will be tempting to conclude that the contractor is paying the price for the items; however, this must be categorised as a cost because the purchased building materials will be used as a production component.

The question to ask related to cost is what will be sacrificed financially to produce these objectives or deliverables to meet the client's needs. From the perspective of the contractor and employer, the cost must be lower for there to be a profit. Therefore, cash flow will be monitored regularly, effective negotiations will be conducted with the subcontractor and overhead costs must not go beyond a benchmark. Variations and changes in scope may lead to additional cost. Consequently, the nature of the contract must be flexible and permissible for compensation when losses arise.

2.2.2 Price

Price, according to the Oxford Learners Dictionary (2021), is "the amount of money that you have to pay for something". The price must never be confused as the cost of a product or service. Price is what is paid to receive a product or service. The price depends on the value of a product or service. In construction activities, the price of an item is set at the point of purchase or rental value. At the end of a building construction, the value of the building is calculated using the yield, internal rate of returns (IRR), or net present value (NPV), and the cost of construction is deducted from the value to provide a price margin for rents or outright purchase.

Price in construction is usually associated with what the end-users will pay for possessing or using the building, road, bridge, or rail. Price is important in understanding the financial returns of a construction project. More importantly, the valuation of a built asset leads to a better understanding of the price that will be set for the end-users to pay.

2.2.3 Value

The Cambridge Dictionary (2021) defined value as "the amount of money that can be received for something". The value of a product or service depends on a number of factors that may raise or lower the final figures attributed to such a product or service. The value of a property depends on the size in terms of cost per meter square, location, life span, interest, construction cost, and ownership. The value of a property may also increase or decrease when the economy in which it is situated rises or is depressed. For instance, in 2008, 2016, and 2020 economic declines, the prices of many properties in affected countries were lower than their original value.

Furthermore, in determining the value of a property, the following formula can be used:

Function / Whole life cost

The function of a property may raise or lower the overall asking price. For instance, a property that will provide monthly financial returns will be valued higher than a property that will not be used for any other function apart from occupying it. The whole life cost of a property considers the investment made into the development of the property and the cost of maintaining the property over a period of years.

In summary, the relationship between cost, price, and value may be summed as follows:

Value − Cost = Price

In simple terms, the price of a property can be determined by deducting the calculated cost of construction from the calculated value. Valuation of a property is a separate topic which is outside the scope of this literature study. The focal point of this chapter is construction cost management nomenclature. Pre-contract costs, post-contract cost control, cost systems, methods, and techniques are discussed in the next subsections.

2.3 Pre-contract cost planning

Pre-contract cost planning or cost control are cost estimating activities conducted before the award of a contract. The aim of pre-contract cost planning is to provide a budgeted cost for construction activities and set out benchmarks for the award of contracts and property valuation purposes. Without pre-contract cost planning, post-contract cost control will be difficult to conduct. Hence, pre-contract cost planning is essential in every project.

Early pre-contract cost control activities started with the quantity surveyors' practice after the Great Fire of London in 1666. Reconstruction activities necessitated a form of measurement and allocation of limited funding. This gave rise to a form of expertise whereby the cost of a project would be known before actual construction. Arguments pertaining to the origin of pre-contract cost planning date back to medieval times. The Bible referred to counting the cost of construction before setting out to construct a tower in Luke 14:28. This implies that there was an ancient common sense accounting system for construction costs before the commencement of any construction activity.

In modern times, the significant pre-contract cost planning activities are commonly identified as being the following:

i   Measurement of construction activities which may be buildings, civil or heavy engineering works
ii Drafting of bill of quantities
iii Rates build up
iv Pricing of bill of quantities
v   Cost planning

vi   Bill of approximate quantities
vii Contractor's estimates of cost
viii Preliminary items of work

Each of these activities is explained in sections 2.6 and 2.7.

2.4 Factors influencing construction estimation

Before commencing cost planning, there are several underlying factors that may limit the veracity of construction cost information. Therefore, the cost planning team must consider the employer's financial limitation, scope, and other expectations in terms of quality, time of delivery, and function of the property. Cost limits must then be set even before designing. This will provide a basis for the project cost control.

Modern construction cost estimation makes use of a spreadsheet and computer-aided design (CAD) measurement. The CAD measurement software facilitates quantification and bill of quantities' preparation with minimal errors. Manual construction cost estimation is prone to errors that are transferred to the tendering and construction processes. As part of the CAD measure, material, labour, plants, equipment, and overhead costs are computed to derive rates which are used to determine the construction cost for each measured item.

There are a number of factors that may influence the accuracy and outcome of a cost estimation process. These factors can be divided into internal and external factors. The internal factors as documented by Ashworth and Perera (2015) are the following:

i   Inconsistent and conflicting design information.
ii   Access to historical cost data affecting the standard of cost planning and quantities produced for the context of the project.
iii Size and complexity of a construction impacting on the nature of errors produced during estimation.
iv   The nature and number of competing contractors working to produce a cost estimate for a construction project. In a keenly contested contract, contractors tend to engage in sub-economical bidding, thereby resulting in inaccurate estimates (if the contract is awarded to a contractor based on lower cost).
v   The quantity surveyors' personal heuristics which are based on experience, qualification, forecasting skills and knowledge.
vi   The experience of the construction project manager and quantity surveyor involved in the estimation process determining the complete cost estimates.

The internal factors mentioned in (i) and (vi) affecting construction cost estimates remains valid for every construction industry in the world. Furthermore, there are some external factors, as discussed by Oyedele (2015), that may invariably impact on the outcome of construction cost estimates:

i Economic conditions resulting from inflation, exchange rate, lending rate, tax rate, import duties and other economic restrictions all have a major influence on the accuracy and outcome of the construction cost.

ii The political situation may encourage higher or lower estimates. In the UK, uncertainties resulting from BREXIT, importation and delivery of materials may lead to additional cost. In developing countries such as Nigeria, the construction estimates are higher in the months leading to the national election. The political situation in a country may encourage or dissuade higher construction cost estimates.

iii Government and regulatory policy affect local content investment, importation, taxation, procurement methods, the inclusion of foreign contractors and standard methods of measurement.

iv The weather conditions of the project location also go a long way in affecting the construction cost. In temperate regions of the world, the autumn and winter seasons are usually avoided, whereas, in tropical countries, the rainy seasons create higher cost due to delays in on-site activities.

v The location of the project also determines the construction cost estimates. In conflict zones such as Iraq, Syria, Niger Delta, and the northern parts of Nigeria, construction cost estimates are usually higher because of insurance cost. The soil conditions on a construction site may also determine the extra cost required for preliminary excavation, road access to the site and setting out phases of construction. For instance, swampy sites will necessitate extra cost for sand filling and other associated preliminary items of work.

vi Construction risks emanating from security issues from the location of the site, sub-contractors' cost, and health epidemics influence the construction cost estimates.

vii Corruption from an external perspective may inflate construction estimates. In cases when kickbacks are required during the tendering and construction phases, the cost of construction will be higher.

Some of the factors mentioned above are avoidable, while the external factors must be mitigated during the course of estimation. Considering the above-mentioned negative factors, thorough examination and evaluation of cost plans and working budgets are fundamental/essential.

It is important to note that construction cost estimation is determined by a standard method of measurement. In the UK, the new rules of measurement (NRM1, 2, and 3) set out the principles of construction estimation. The NRM1 provides guidance on the levels of cost planning, the NRM2 provides descriptions and notes on measurement of building works, while the NRM3 stipulates guidance on the order of cost estimates and building maintenance. Civil engineering works are guided by the Civil Engineering Standard Methods of Measurement (CESMM). In other countries globally, construction estimation is supported by estimation standards which are established by their institutes of quantity surveying.

***

PART II

2.5 Construction cost management system

The Cambridge Dictionary (2021) defined a system as a set of interconnected things that work together. The Business Dictionary (2021) also defined a system as a set of methods, procedures, and routines created to carry out specific activities or solve a problem. A system consists of interconnected input, output, and feedback loop methods, processes and boundaries for self-evaluation (Hoyle, 2009).

Construction cost management systems act as umbrellas for a broad range of methods and techniques which can be deployed for cost management, for example, the value management system encompassing value analysis, planning engineering. The next subsection identifies and explains all applicable construction cost management systems.

2.5.1 Traditional cost management system (TCMS)

The traditional cost management system has been used since the inception of the quantity surveying profession when civil and building engineering works required a measurer to produce broken down quantities necessary to complete a work. Since 1836, the quantity surveying profession has gradually emerged and made use of cost planning and controlling techniques such as taking-off, cash flow registers, and the bills of quantities.

The traditional cost management system has been used in many countries globally. This system has since evolved using digital tools such as Microsoft Word and Excel Spreadsheet. Modern cost databases such as the Building Cost Information Service (BCIS) in the UK have afforded quantity surveyors and construction stakeholders the opportunity to reuse historical data for cost estimation and budgeting.

Table 2.1 identifies the components and users of the traditional cost management system, its strengths and weaknesses. The main users of documents produced from the traditional costing process are quantity surveyors, contractors, estimators, cost consultants, clients, contractors and subcontractors, and suppliers.

The traditional cost management system has been criticised for its lack of innovation in the changing times. Koskela et al. (2002) supported this

Table 2.1 Features of traditional cost management system

Cost management system	Components	Tools	Strengths	Weaknesses
Traditional cost management	Cost estimation Cost budgeting, taking-off Bills of quantities Cost control and monitoring Life cycle costing	Cost database: BCIS Cost estimation digital tools CAD measurement tools MS Word and Excel Spreadsheet Cash flow register	Suitable for simple construction projects Cost planning and controlling tools	Common problems are cost and time overrun Inaccurate costs Estimating challenges Confusing cost management system Not suitable for complex and modern projects

claim by stating that the traditional cost management system does not deliver the expected productivity when compared to other forms of construction cost management systems. Furthermore, an integrated approach is required for enshrining product and system cost estimation requirements for effective decision-making in complex projects.

2.5.2 *Value management system (VMS)*

Value management is a system whereby a balance is created between cost, time, quality, and risk to deliver value for money (Kineber, Othman, Oke, Chileshe and Buniya, 2021; Madushika, Perera, Ekanayake and Shen, 2020). On construction projects, value management seeks to choose between alternatives to decipher tasks that will deliver the best quality at the lowest cost. Hence, the essence of value management as a process starts with planning, execution, and performance measurement. Consequently, the main components of value management are value planning, engineering, and analysis. The aforementioned trio are construction cost management methods which are explained under Section 2.3. Value management engages with all relevant stakeholders in a structured, methodical, and concise approach to create more value in the construction project.

As stated in Table 2.2, the main demerits of adopting value management relate to the meticulous nature of the processes involved in value management and the requirement for representation from all key stakeholders in the construction project. The process of conducting value management is time-consuming and in some specific procurement approaches, may not be feasible. Nevertheless, value management delivers high-quality construction products with the aid of key performance indicators (KPIs).

Table 2.2 Feature of value management

Cost management system	Components	Tools	Strengths	Weaknesses
Value management (VM)	Value planning Value engineering Value analysis	Brainstorming tools MS Excel Spreadsheet Construction cost data such as BCIS	Starts from the planning phase of the project up until the completion. Involves stakeholders in reducing cost, improved value, and better resource allocation	Cost of the study, involving a broad representation of stakeholders in the study, to be combined with other cost management systems such as life-cycle costing. The scope can change often

Tools used in delivering value management include brainstorming techniques and digital tools such as Microsoft Excel and construction databases such as the BCIS.

2.5.3 Earned value management system

The earned value management system is an overarching approach for measuring and forecasting the cost performance of a construction project (Lipke, Zwikael, Henderson and Anbari, 2009). This management system integrates the schedule, cost, and value of a construction project in the evaluation process (Sutrisna, Pellicer, Torres-Machi and Picornell, 2020). It is important to note that the earned value management system should not be confused with the value management system. An earned value management system is different from a value management system in terms of multiple components used to compute the outcomes. A good understanding of the components of the earned value management system was summarised as follows by Lipke *et al.* (2009, p. 401):

> Planned value (PV); actual cost (AC); earned value (EV); cost variance (CV = EV − AC); schedule variance (SV = EV − PV); cost performance index (CPI = EV/AC); schedule performance index (SPI = EV/PV); budget at completion (BAC, the planned cost of the project); performance measurement baseline (PMB, the cumulative PV over time); independent estimate at completion (IEAC, the forecasted final cost).

Table 2.3 provides a quick overview of the constituents of the earned value management system. The most useful digital tool for producing the earned value management components is the Microsoft Office Project.

Table 2.3 Features of the earned value management system

Cost management system	Components	Tools	Strengths	Weaknesses
EVMS	EV analysis CV BAC CPI Estimate at completion (EAC) Estimate to completion (ETC) To complete at performance index (TCPI) IEAC	MS Project and portfolio management Primavera software Earned value software EVMax and associated software for EVMS MS Excel Spreadsheet	Starts from the planning phase throughout the project, used for forecasting Integrates time, cost, and scope Gives more accurate estimates for complex projects	Involves a great deal of calculations from a large amount of data Depends on available knowledge of software packages Output depends on application of software and data manipulation

Other project management software can also be used to track and measure project progress and performance. The accuracy of calculations produced during the cost evaluation phases is the main merit of applying this construction cost management system. However, it will be impossible to conduct earned value management in the absence of a trained professional who is skilled in making use of the digital tools for measuring and tracking project cost, schedule, and value.

2.5.4 5D Building information modelling management (BIM)

Bryde, Broquetas and Volm (2013, p. 971) defined building information modelling (BIM) as:

> a set of interacting policies, processes and technologies generating a methodology to manage the essential building design and project data in digital format throughout the building's life-cycle.

Building information modelling has evolved from being a methodology for building design, data management and building life to a management system for delivering construction projects into a cost management framework. BIM as a construction cost management system also produces cost analysis and plans according to the levels of detail (LOD) of the design. The 5D BIM process involves an iteration of cost analysis through the planning, construction, and in-use phases. This leads to the development of the BIM execution plan (BEP) and Construction Operations Building Information Exchange (COBIE) data, where construction cost is evaluated during construction in BEP and the life cycle cost is applied through the COBIE data.

Table 2.4 Constituents of building information modelling (5D BIM)

Cost management system	Components	Tools	Strengths	Weaknesses
5D BIM	Cost analysis and planning Employer or exchange information requirement Master information delivery plan (MIDP) BIM execution plan (BEP) COBIE data	MS Excel Spreadsheet Digital software applications such as Revit, Cost X, Bluebeam	Improved communication, easier cost management from the design, improved profit, quality, and client satisfaction Ensures sustainability in construction Effective cost monitoring and improvement Extends to the in-use and operational stages of a facility	Expensive for most construction companies in developing countries because of the cost of software. Requires training and retraining of staff

Table 2.4 highlights the essence of BIM as a management tool, the cost components produced using relevant digital tools such as Cost X, Revit, and other CAD measure software applications. BIM ensures better quality communication, simpler cost management from designs, and improved profit, quality, and client satisfaction. In many developing construction economies, BIM has been touted as being expensive. The expenses were documented as software licenses, regular training, and the scale of the project. In 2011, the UK government produced a mandate for BIM level 2 adoption in all public works by the year 2016 (Georgiadou, 2019). The concept of 5D BIM emanated from the 5th dimension of BIM. BIM 3D involves 3D designs, 4D BIM incorporates schedule information with the 3D design, and 5D BIM includes the cost information.

7D and 8D BIM embrace sustainability and asset safety management, respectively. In 5D BIM, cost plans are developed in accordance with the NRM1, 2, and 3 (2012). The NRM were developed by the Royal Institute of Chartered Surveyors (RICS) in the year 2012. The NRM1 focuses on the order of cost estimate and cost plans 1, 2, and 3, whereby early cost advice is given to the client in cost plan 1. Cost plans 2 and 3 provide a more detailed cost plan as the drawing details emerge. NRM2 contains a structured approach for measuring quantities and preparing the bill of quantities, while NRM3 details the operation and maintenance cost for life cycle costing. Hence, in 5D BIM, the overall process of the traditional cost management system is not only incorporated but also enhanced with

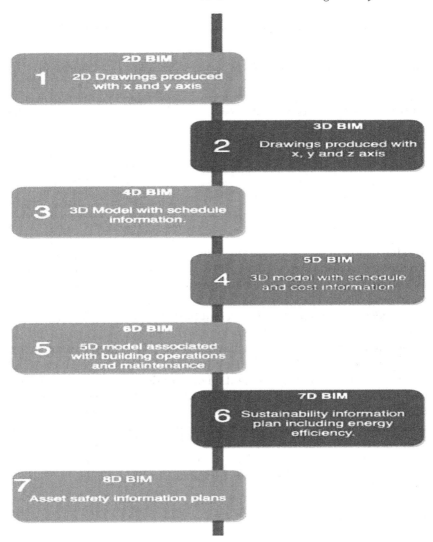

Figure 2.2 BIM process.

Source: Authors.

the use of digital tools for CAD measurement and the collaboration of all relevant stakeholders in the cost plan and budgeting development.

2.5.5 Benchmarking system

The benchmarking costing system became popular in the early 2000s in the United States. Benchmarking is mostly synonymous with productivity and quality improvements (Park, Thomas and Tucker, 2005). Cost

Table 2.5 Features of cost benchmarking

Cost management system	Components	Tools	Strengths	Weaknesses
Benchmarking system	KPIs Average cost Historical cost data and baselines	MS Project MS Excel Cost databases (e.g. BCIS)	More value for money, improved quality Waste reduction during the construction process is also achieved	Requires huge database which has to be updated regularly and adequate industry analysis to establish KPIs

benchmarking in the construction industry is based on average cost and baselines from historical cost data. KPIs are used to measure outline targets for cost performance. The system involves adopting a general approach for measuring construction cost and value by gathering cost data from previous projects and comparing the present performance of a particular project (s) to the industry's best practices in terms of cost.

The merit of using cost benchmarking is the creation of value for money, as indicated in Table 2.5. Cost benchmarking is based on creating more value in labour, material, and quality productivity. The weakness of cost benchmarking is that a large data set is required to determine the cost benchmark. Databases such as the BCIS in the UK are very reliable for cost benchmarking. However, in some developing countries where there are no construction cost databases, it is difficult to implement cost benchmarking. Notwithstanding cost benchmarking can be used along with traditional cost management where historical cost data is readily available.

2.5.6 Expert systems

The aim of an expert system is to bypass the weaknesses of conventional construction cost management systems. Park *et al.* (2005) developed a framework for a cost management expert system (COMEX) for the sole purpose of enhancing the decision-making process in construction cost management. The development of an expert system depends on pre-existing knowledge; therefore, knowledge management and engineering are the bedrock of construction cost expert systems. Value for money is achieved in expert systems such as COMEX by adopting the resources with the lowest cost as a cost-saving measure. The expert system is also known as expert judgement and involves both algorithmic and non-algorithmic cost models. The expert system of construction cost management consists of digital software applications designed to satisfy clients' requirements by achieving

Table 2.6 Expert systems components, tools strengths, and weaknesses

Cost management system	Components	Tools	Strengths	Weaknesses
Expert system	Knowledge management and engineering Activity-based costing (ABC) Cost minimisation models Maintenance cost plans	Digital software application such as COMEX ROSE Surface Condition Expert System for Pavement Rehabilitation (SCEPTRE) PERSERVER ERASME EXPEAR PAVEMENT EXPERT PARES PAVER and Micro PAVER AIRPACS	Can be used by non-cost estimators Software-based and easier to use for every other type of project other than building works Incorporates facilities management plan Reduces cost	May be inaccurate Expensive specialist software Technical know-how is required to adopt an expert system software application

value for money and reducing cost. The specialist software applications are not only designed for building construction but also for any other project. It may be developed for pavement construction or the maintenance and operation of a facility.

There are several software applications that have been designed for the purpose of cost minimisation as indicated in Table 2.6. The main weakness of expert systems is their level of accuracy: not all expert systems provide accurate results. Some of these software applications may be designed for a local government, ministry, or government agency. Therefore, the needs of these software applications are only tailored to the specific needs of a client's project and will provide a streamlined output. The cost of creating these applications for a specific client may be on the high side when compared to other forms of construction cost management. Furthermore, regular training and updated cost data are required to facilitate the output of the required cost through expert systems software. The main benefit of expert systems is evident in their unique cost minimisation output as opposed to a traditional cost management system.

2.6 Summary

For a thorough understanding of continuous cost improvement as a new method for planning and managing construction cost, construction cost management must be understood. Therefore, construction cost systems, methods, and techniques present a structured approach to comprehensively planning, evaluating, applying, and managing construction cost. The array of construction cost management options available to a construction planning team, with emphasis on the cost planning professionals, must guide the user towards innovative construction cost management systems, methods, and techniques such as continuous cost improvement.

References

Ashworth, A. and Perera, S. (2015). *Cost studies of buildings* (6th edition). Routledge: London.

Bryde, D., Broquetas, M. and Volm, J. M. (2013). The project benefits of building information modelling (BIM). *International Journal of Project Management*, 31(7), pp. 971–980. https://doi.org/10.1016/j.ijproman.2012.12.001.

Cambridge Dictionary. (2021). Available from: https://dictionary.cambridge.org/dictionary/english/cost (Accessed 8, February 2021).

Georgiadou, M. C. (2019). An overview of benefits and challenges of building information modelling (BIM) adoption in UK residential projects. *Construction Innovation*, 19(3), pp. 298–320. doi: 10.1108/CI-04-2017-0030.

Hoyle, D. (2009). ISO 9000 Quality Systems Handbook-updated for the ISO 9001: 2008 standard.

Jainendrakumar, T. D. (2015). Project human resource management for project managers based on the PMBOK. *PM World Journal*, 4(VIII), pp. 1–14.

Kern, A. P. and Formoso, C. T. (2004). Guidelines for improving cost management in fast, complex and uncertain construction projects. In *12th Conference of the International Group for Lean Construction*, Helsingor, Denmark (pp. 220–233).

Kineber, A. F., Othman, I., Oke, A. E., Chileshe, N. and Buniya, M. K. (2021). Impact of value management on building projects success: Structural equation modeling approach. *Journal of Construction Engineering and Management*, 147(4), p. 04021011. doi: 10.1061/(asce)co.1943-7862.0002026.

Koskela, L., Howell, G., Ballard, G., and Tommelein, I. (2002). The Foundation of Lean, in *Design and construction: Building in value*, pp. 211–226.

Lipke, W., Zwikael, O., Henderson, K. and Anbari, F. (2009). Prediction of project outcome: The application of statistical methods to earned value management and earned schedule performance indexes. *International Journal of Project Management*, 27(4), pp. 400–407. https://doi.org/10.1016/j.ijproman.2008.02.009.

Madushika, W. H. S., Perera, B. A. S., Ekanayake, B. J. and Shen, G. Q. P. (2020). Key performance indicators of value management in the Sri Lankan construction industry. *International Journal of Construction Management*, 20(2), pp. 157–168. doi: 10.1080/15623599.2018.1484556.

Owens, J., Burke, S., Krynovich, M. and Mance, D. J. (2007). Project Cost Control Tools and Techniques. [Online]. Available from: https://pdf4pro.com/view/project-cost-control-tools-amp-techniques-jason-3c20.html

Oxford Learners Dictionary. (2021). Available from https://www.oxfordlearnersdictionaries.com/definition/english/price_1?q=price (Accessed 8, February 2021).

Oyedele, O. A. (2015). Evaluation of factors affecting construction cost estimation methods in Nigeria. Conference: FIG Working Week 2015 From the Wisdom of the Ages to the Challenges of the Modern World Sofia, Bulgaria, 17–21 May 2015. Sofia, Bulgaria.

Park, H. S., Thomas, S. R., and Tucker, R. L. (2005). Benchmarking of construction productivity. *Journal of Construction Engineering and Management*, 131(7), pp. 772–778. doi: 10.1061/(asce)0733-9364(2005)131:7(772).

Rad, P. F. (2002). Project estimating and cost management. Management Concepts, Vienna, Virginia.

Royal Institution of Chartered Surveyors (RICS) (2012). *RICS new rules of measurement* [Online]. Available from: https://www.rics.org/uk/upholding-professional-standards/sector-standards/construction/rics-nrm-new-rules-of-measurement/ (Accessed 4, March 2021).

Sutrisna, M., Pellicer, E., Torres-Machi, C. and Picornell, M. (2020). Exploring earned value management in the Spanish construction industry as a pathway to competitive advantage. *International Journal of Construction Management*, 20(1), pp. 1–12. doi: 10.1080/15623599.2018.1459155.

The Business Dictionary. (2021). System [online]. Available from: https://www.dictionary.com/browse/system (Accessed 3, February 2021).

Tichacek, R. L. (2006). Effective cost management-back to basics. *Cost Engineering*, 48(3), p. 27.

Venkataraman, R. R. and Pinto, J. K. (2011). *Cost and value management in projects*. John Wiley & Sons: Oxford, UK.

3 Construction Cost Management Methods

Temitope Omotayo, Udayangani Kulatunga, and Bankole Awuzie

3.1 Introduction: Construction cost management methods

Construction cost management methods are distinctly dissociated from construction cost management systems. The nomenclature of "method" can be described as a means of procedures, especially a procedural way of fulfilling an endeavour (Free Dictionary, 2021). The Merriam-Webster Dictionary (2021) defines method as a technique of doing something. A method can be an established habitual, logical, or prescribed practice of achieving specific ends with accuracy and efficiency, usually in an ordered sequence of fixed steps (Cambridge Dictionary, 2021). These definitions provide a clear delineation between what systems and methods are. Construction cost management methods are drawn from construction cost management systems.

For instance, construction cost management methods such as cost estimation and material take-off are drawn from the traditional cost management system. Other construction cost management methods include value planning, engineering, and analysis born out of the value management system. Basically, these methods are components of construction cost management systems required to deliver the systems' objectives. Similarly, these construction cost management methods have their own internal components known as techniques. Subsequent sections explain and describe the components of construction cost management methods.

3.2 Cost estimation and material taking-off

One of the main methods used to deliver the traditional cost management system is cost estimation and material taking-off. Construction cost estimation is a method employed to detail construction elements, activities, calculated costs, and allocation of costs. In the UK, the NRM1 document suit provides guidance on the stages of cost estimation. NRM1 provides guidance on how cost planners can provide initial cost advise by using historical cost databases such as the BCIS for determining early cost before detailed designs.

DOI: 10.1201/9781003176077-4

The basic parameters used in cost estimation as defined by RICS (2012) are the following:

   i Gross internal floor area (GIFA)
  ii Cost per meters squared (cost/m²)
 iii Cost per square metre of gross internal floor area (cost/m² of GIF)
 iv Elemental unity quantity (EUQ)
  v Elemental unit rate (EUR)
 vi Elemental cost (EC)
vii Gross internal floor area per metres squared (GIFA/m²)

According to RICS (2012), the gross internal floor area (GIFA) is the total area of the internal floor, including the internal walls. When the internal walls are excluded, it becomes the net floor area (NFA). In most calculations involving early cost advice and other estimations, the GIFA is multiplied by the cost per metres squared to produce the total cost of the building.

Other calculations to deliver the estimated cost of a building can be summarised as follows:

$$Elemental\ cost(EC) = Elemental\ unit\ quantity(EUQ) \times Elemental\ unit\ rate(EUR)$$

$$(3.1)$$

Elemental cost $(EC) =$

Cost per square metre of gross internal floor area $\left(Cost/m^2 of\ GIF\right) \times$ (3.2)
Elemental unit rate (EUR)

Elemental costs are compiled in a cost plan spreadsheet format to produce the total construction cost. The first cost plan is usually inaccurate and only provides an idea of what the project cost will look like. Additional costs from external works, plumbing, electrical installations, gas, mechanical services, preliminary items of works, marketing cost, design team feeds, contingency percentage, inflation, and tax rates are included in the cost plan. Cost plan 3 is adjusted from the second cost plan once a bill of quantity has been produced.

The bill of quantities (BOQ) document is an organised structured budget produced from a process of measuring the exact quantities of materials from a working drawing, abstracting the quantities, and preparing concise documentation of the cost of activities as taken from the rate required to complete each activity. The taking-off process is the first method employed in creating the BOQ. An abstracting process involving manual or computational deductions and additions of the measured quantities (which may be in metres, metres squared, metres cubed, or numbers) will follow before imputing the quantities into a BOQ format.

Table 3.1 Constituents of cost estimation and material taking-off

Cost management method	Components	Tools	Strengths	Weaknesses
Cost estimating and material taking-off	Gross internal floor area (GIFA) Cost per meters squared (£/m²) Gross internal floor area per meters squared (GIFA/m²) Elemental unit quantities (EUQ) Elemental unit rate (EUR) Elemental cost (EC) Cost plans 1, 2, and 3 Bills of quantities Cost monitoring and control during execution	New rule of measurement 1, 2, and 3 (NRM 1,2, 3) CAD Measurement software such as Cost X, Bluebeam, quantity take-off MS Excel and MS Word	Provides the required data for the creation of tender documents Provides a guideline for construction cost control and monitoring	Easily subject to cost and time overrun. This method can contain latent errors All problems associated with traditional cost management systems are evident in the cost estimation and material taking-off process

The rate, which is the unit cost required to complete each activity of work in the BOQ, can be calculated by compiling the material, labour, plant, or equipment cost along with allowance for wastage which is usually taken as a percentage. Each rate produced for the work activities in the BOQ is multiplied by the quantities to provide the total cost.

The BOQ is an output of the cost estimation and taking-off process (as shown in Table 3.1). BOQs are considered to be part of the tender documents contractors must price and submit as part of the project bidding process. As a working budget for the construction process, the BOQ ensures cost monitoring and control. However, the lapses associated with traditional costing have been evident in the incidences of cost overruns, variations, errors in the BOQs, and construction disputes prevalent in instances where this method has been deployed. Cost estimation and taking-off methods have been in existence since the onset of construction cost management. Innovations in construction cost management have not left out the components of cost estimation and taking-off. This method is the most commonly used in construction cost management and the combination of CAD Measure software application has to some extent mitigated the drawbacks of traditional costing and cost estimation methods.

3.3 Value planning

Value planning is also a method under the value management system. Value planning is the first phase of value management, whereby economic returns are projected from the client's brief, working drawings, cost information, and weighted value criteria to deliver assets that are environmentally friendly at the best quality and cost (Gayani, Uthpala & Dushan, 2016). Gayani et al. (2016) indicated that value planning can be combined with sustainability concepts to produce built assets of a higher quality. Although value planning has been combined with other cost management methods, the essence of value planning in construction cost management is to create built assets imbued with a higher quality but at a lower cost. Hence, clients' requirements are the basis of value planning. Value planning entails decisions that affect the project and construction stakeholders to deliver economically, environmentally, and socially friendly properties. Value planning aids the development of material quality options for the construction process.

Table 3.2 Value planning, components, tools, strengths, and weaknesses

Cost management system	Components	Tools	Strength	Weaknesses
Value planning	Preferred scheme Strategic policies	Client's brief Brainstorming Working drawings Cost data Evaluation Weighted value criteria	Policies are created during the value planning phase to reduce cost and have a positive impact on the construction companies as a whole	Just like the value management system, value planning is cost- and time-consuming

The process of creating a list of alternatives for value creation in construction projects involves multiple options of cost, quality, economic, and sustainability data. As indicated in Table 3.2, value planning depends on enormous amounts of construction data that are not readily available. The time taken to produce a value plan in a project is one of the discouraging factors of applying this cost management method. Furthermore, value planning cannot be used alone without combining it with other construction cost management methods such as cost estimation, value analysis, or engineering.

3.4 Value engineering

Value engineering can be traced back to the end of the Second World War. The application of value engineering as a construction method is more popular in the United States, the United Kingdom, Europe, and more recently in Asia. Value engineering is the second stage of a value management system. Value engineering has been described as the construction phase of value management whereby the development of various options during the construction stage influences changes in the quality of the built asset produced. The overall aim of value engineering is to create more value in the construction process. Cheah and Ting (2005) opined that value engineering can be closely associated with target costing by creating life cycle costing with well-defined requirements and quality. The process of value engineering must lead to the desired level of profitability throughout the life cycle of a built asset (Cheah & Ting, 2005). Thus, the process of establishing value during construction depends on clients' requirements, clients' support and active participation in the value management team, the qualified value management facilitator, and a brainstorming workshop (Shen & Liu, 2003). Value engineering depends on information gathered during the construction process, cost-value reconciliation and evaluation, and regular site meeting and appraisals. The agenda of value engineering is to identify current alternatives such as building materials and construction processes that may be more sustainable and deliver higher quality at a lower cost for quality delivery of the project. Client satisfaction is always on the agenda of the value engineer.

The outcome of the value engineering process depends on the quality of information gathered during the construction process. It is essential to note that value engineering must not conflict with the client's original requirement for the purpose of avoiding variations. Table 3.3 depicts the main merits of adopting value engineering along with methods such as target costing for waste minimisation and profit enhancement. However, the demerit may likely be time overrun if the value engineering sessions are not properly managed.

Table 3.3 Value engineering components, strengths, and weaknesses

Cost management method	Components	Tools	Strengths	Weaknesses
Value engineering	Value engineering reports Functional analysis Implementation and follow-up tools	Information gathering Value and cost evaluation Site meeting and appraisals	Waste reduction, improved profit for the company, and improved competitiveness	The cost and time of investigating various options for improved value can lead to cost and time overrun if not adequately planned

3.5 Value analysis

Value analysis stems from the value management system. Value analysis is applicable in the operation and maintenance of existing buildings. The core objective of value analysis is to improve the functionality of a property. Cariaga, El-Diraby, and Osman (2007) described value analysis as the process of evaluating the functional requirements, equipment, facilities, services, and supplies in an asset in a systematic manner to reduce pointless costs. In the process the quality, performance, and safety of the asset are maintained to the client's satisfaction. Value analysis examines the functionality of a property in terms of quality, value for money, and client satisfaction.

Operational and maintenance costs derived from value analysis are produced during the design and construction stages. In the design stage of a property, the value of the functional options is evaluated in terms of quality and cost. A form of a quality function deployment technique is integrated with client requirements and a decision-support mechanism is arrived at when finalising the technical working designs. The value analysis offers benefits in terms of cost minimisation of operations and the maintenance of properties, producing higher-quality assets. The main weakness of value analysis identified in Table 3.4 is its dependence on the whole life cycle costing process, whereby cost data associated with operations and maintenance must also be used. This process is quite daunting and involves intense brainstorming with pertinent stakeholders for research purposes.

Table 3.4 Features of value analysis

Cost management system	Components	Tools	Strengths	Weaknesses
Value analysis	Quality function deployment (QFD) Correction of defects Operation and maintenance cost analysis	Brainstorming tools MS Excel Process monitoring Feedback tools	Value analysis is also used in cost reduction activities This also affects the value of the building in the long term	The basic challenge with value analysis is that it has to depend on the whole-life cycle costing data and other stakeholders for the research

3.6 Whole life cycle costing

Whole life cycle costing compares the opportunity and future cost of a building by involving investment cost and income flow within the project. Kishk, Al-Hajj, Pollock, Aouad, Bakis, and Sun (2003) noted that whole life cycle costing started in the 1950s when the Building Research Establishment (BRE) created the concept of cost-in-use. In 1977, a guideline on how whole life cycle costing can be applied in the construction industry was released by the department of industry. Whole life cycle costing is a costing method that facilitates the evaluation of cost performance during construction and the in-use phases of properties through the consideration of several alternatives (Kishk et al., 2003). The construction best practice programme (CBPP, 1998) defined whole life cycle costing as follows:

> The systematic consideration of all relevant costs and revenues associated with the acquisition and ownership of an asset.

The process of calculating whole life cycle costing entails consideration of all associated costs, ranging from land acquisition, cost of construction, revenue generated during the operational phase of the facility, as well as other related costs. Generally, whole life cycle costing makes use of mathematical models which depend on the future cost of money. Therefore, formulas associated with the net present value (NPV), basic interest formula, present worth of annuity, single present worth, equivalent annual cost (EAC), discounted payback period (DPP), and the internal rate of return (IRR) are some of the important tools used in whole life cycle costing decision-making.

Generally, the components of whole life cycle costing are contained in its NPV formula:

$$NPV = C + R - S + A + M + E \qquad (3.3)$$

Where C = cost of investment, which includes land acquisition, construction, marketing, and professional fees

R = Replacement cost during execution
S = Resale value at the end of the study period
A = Annually recurring operating, maintenance, and repair costs (except energy costs)
M = Non-annually recurring operating, maintenance, and repair costs (except energy costs)
E = Energy cost

The level of data required for whole life cycle costing calculation can be structured around the project, phase, category, element, and task levels (El-Haram, Marenjak & Horner, 2002). The project level consists of the overall project costs. The phase level breaks down the cost at each construction phase; the categorical level breaks down the cost into an elemental format; the elemental level provides cost breakdown in the form of tasks costs; and finally, the tasks level consists of the resources and activity costs.

Table 3.5 Features of whole life cycle costing

Cost management method	Components	Tools	Strengths	Weaknesses
Whole life cycle costing (WLC)	Investment costs Replacement costs The resale value at end of study period Annually recurring operating, maintenance, and repair costs (except energy costs) Non-annually recurring operating, maintenance, and repair costs (except energy costs) Energy costs	Net present value (NPV) Basic interest formula Present worth of annuity Single present worth Equivalent annual cost (EAC) Discounted payback period (DPP) Internal rate of return (IRR)	Very useful for investment appraisal Reliable analysis	It is mostly confused with life-cycle costing There is an early struggle for profitability, and there is a drop in productivity Paying back the loans for the investment can be a challenge

The benefits of whole life cycle costing, as indicated in Table 3.5, are provided in terms of clear investment appraisal and reliable analytical costing method. Conversely, the demerits are its focus on profitability instead of value or productivity. Whole life cycle costing can also be confused with life cycle costing.

3.7 Life cycle costing

Life cycle costing should not be misconstrued as whole life cycle costing. The difference between whole life cycle costing and life cycle costing revolves around the stages of application. Unlike whole life cycle costing that considers all costs, including land acquisition and construction, life cycle costing only considers operational and maintenance costs by comparing alternative costs, conducted through the planning and construction stages of the construction project (Cole & Sterner, 2000). Life cycle costing was developed by the US Department of Defense as a costing decision-making tool for the procurement of military equipment in the mid-1960s (Cole & Sterner, 2000). Life cycle costing is not a popular construction costing method because it requires access to large quantities of historical and reliable building cost data.

The formula for calculating life cycle costing is given as follows:

$$LCC = C + PV \ Recurring - PV \ Residual \ Value \tag{3.4}$$

Table 3.6 Features of life cycle costing

Cost management method	Components	Tools	Strengths	Weaknesses
Life cycle costing	Construction cost Recurring cost Residual value of a property	MS Excel Digital databases Building cost information service (BCIS)	Includes preliminary item of works during construction Running cost, energy bills, utilities cost used before, during, and after construction	Can be expensive to use, and has to be combined with value management for it to be effective Difficulties accessing digital software and database

Where C = construction cost at year 0

LCC = Life cycle cost
PV = Present value of all recurring cost
PV = Present value of residual value at the end of the project

The advantage of using life cycle costing during construction for cost management is its ability to encapsulate the preliminary items of work and associated costs as part of the construction cost. The total construction cost, residual, and recurrent costs in terms of present value are all calculated over a period of 60 years or more. Table 3.6 also noted that the weakness of using life cycle costing is the expensive nature of the process of software. Life cycle cost depends on historical and current cost data. The calculations may become complex in larger projects. Further life cycle costing is mostly used for the in-use stage of a property, while other construction cost management methods such as value engineering and traditional costing methods may be combined with life cycle costing.

3.8 Activity-based costing

Activity-based costing (ABC), just like other construction cost management methods, started after the Second World War. The ABC method became prominent in the 1990s as an on-site construction cost management method (Lin, Collins & Su, 2001). ABC in the construction industry is an accounting method that considers the individual cost of activities and such costs allocated to products and services on the basis of activities to deliver each product or service (Lin et al., 2001). In ABC, construction cost drivers such as overheads are identified and mitigated. The ABC method

Table 3.7 Features of activity-based costing

Cost management method	Components	Tools	Strengths	Weaknesses
Activity-based costing	Interim valuations Milestone targets	Invoices indicating expenses Labour, plant, and equipment cost Preliminary costs MS Excel Spreadsheet	Used during construction. The very effect in reducing overhead and services cost Integrated with supply chain	May be more effective to combine with traditional costing and or target costing Detailed cost data from the construction process is required

depends on expenses incurred during construction and there is an integration with the supply chain management. The cost of the supply chain has to be measured for an effective ABC method during construction. In some instances, milestones are set in order to ensure the effective implementation of ABC. Interim valuations can be used as a technique for measuring ABC at regular intervals (as stated in Table 3.7).

ABC is a construction cost controlling method which is dependent on the level of cost information gathering. This may be a merit or demerit depending on the level of cost data available. Consequently, during the planning phases of a construction project, a framework for implementing ABC must be established.

3.9 Earned value analysis

Earned value analysis (EVA) stems from the earned value management system. EVA measures the performance of project cost and schedule in terms of established baselines of the planned scheduled and actual cost. EVA measures work in progress in terms of cost, time, and value (Leu & Lin, 2008). EVA makes use of the planned and actual value of construction works to provide a final value in the form of a cost variance (Ankur & Pathak, 2014). The output of EVA provides a form of forecasting showing plausible outcomes of the project cost and schedule. The main parameters for measuring the EVA are the cost variance, schedule variance, cost of performance indices, schedule performance indices, estimate to completion, and variance at completion (Virle & Mhaske, 2013). EVA is conducted with the aid of software applications such as Microsoft Project, Asta Powerproject, Primavera, and Microsoft Excel Spreadsheet.

EVA provides an opportunity for accurate cost forecasting and cost savings. However, as indicated in Table 3.8, regular training and technical skills are required on the part of the cost consultant. It is also important

Table 3.8 Features of earned value analysis

Cost management method	Components	Tools	Strengths	Weaknesses
Earned value analysis	Planned value Actual cost Variance analysis Schedule variance Cost variance Performance indices Estimate to completion Variance at completion	MS Project and portfolio management Asta Powerproject Primavera software Earned value software EVMax and associated software for EVMS MS Excel Spreadsheet	Aids organisation of work activities into work breakdown structures (WBS) Measures value and performance A modern and effective method for project cost control	Cost consultant must be skilled in using relevant software May be misused for forecasting

to note that some of the forecasts in EVA may be inaccurate if there are errors in some of the parameters. This may result in early cost overruns, especially in complex construction projects.

3.10 Target value design and costing

Target value design and costing methods emanate from the benchmarking cost management system. Target value design started in the manufacturing sector in the United States in the 1930s before it emerged in the construction sector in the 1960s (Zimina, Ballard & Pasquire, 2012) with the Boldt construction in St. Olaf's Tostrud Fieldhouse and Thedacare's Shawano Clinic projects (Pishdad-Bozorgi & Gao, 2018). Target value design combined target costing and value management in design. Target costing and value management planning can be combined during project design to produce high-quality design and cost plans. Target costing is aimed at creating more value during the planning and construction phases of a project. Target costing became popular in the Japanese manufacturing sector after the Second World War (Zimina et al., 2012). Target costing planning aids the minimisation of project costs by reducing unnecessary expenses and ensuring that the best quality elements are incorporated in the design.

The tools used in target value design are dashboard and huddle boards, cluster teams, computer visualisation and simulation, and physical prototypes (Alves, Lichtig & Rybkowski, 2017). The essence of these tools is to create collaboration and engage all stakeholders in the design

Table 3.9 Features of target value design and costing

Cost management method	Components	Tools	Strengths	Weaknesses
Target value design and costing	Target costing planning Target cost contracts	Dashboard and huddle boards Cluster teams Computer visualisation and simulation Physical prototypes	Waste reduction and cost minimisation Improved profitability Combined target costing and value engineering during design for easier cost control during construction	Cannot be used alone. It has to be combined with value management Mostly applicable in the design phase

and cost planning phases. Target value design and costing depend on a well-defined client's brief, basic net price, pre-calculations, direct cost, and feasibility studies.

Cost limits and targets are associated with all elements of the building and the design is produced according to the cost limits and targets. The major demerit of target value design and costing is its dependence on other methods and systems for its effectiveness, as stated in Table 3.9. Target value design and costing is a lean thinking concept that is based on reducing construction physical and non-physical waste. The result of a well-implemented target value design and costing method in construction is increased profitability for the contractor, a high-quality built asset, and client satisfaction.

3.11 Kaizen costing

Kaizen costing is the cost management aspect of kaizen, which is a Japanese word for continuous improvement (Omotayo, Awuzie, Egbelakin & Obi, 2020). Kaizen costing has been applied in the manufacturing industry as a lean thinking concept over the years. However, its relevance in the construction industry is still in doubt because of the level of awareness amongst construction professionals. On the other hand, Vivan, Ortiz, and Paliari (2015) demonstrated the application of kaizen costing in the construction process of buildings in Brazil. Their findings confirmed previous literary findings supporting continual cost reduction through waste minimisation for higher quality and client satisfaction. Kaizen costing operates by continually identifying physical and non-physical waste during construction for the purpose of mitigation.

Omotayo and Kulatunga (2015) noted that kaizen costing can be implemented during the construction under the categories of improvement and maintenance of cost. Overhead costs associated with project

Table 3.10 Features of kaizen costing

Cost management system	Components	Tools	Strengths	Weaknesses
Kaizen costing	Cost limits identification Overhead cost management Work breakdown structure	Plan-do-check-act principles (Deming cycle) Kaizen costing register	Generates more profit for the contractor, improved quality, and satisfaction	Mostly applicable in the construction phase Training and experience in the use of this method required

labour, plant, equipment, material purchase orders, subcontractors' payment, variation, and preliminary items of work are incrementally reduced but not eliminated. Table 3.10 indicates that overhead cost is an integral part of every cost. Construction cost maintenance involves a management function targeting narrowed down policies and guidance on waste reduction in the office, encouragement of employee-employer relationship and waste elimination. The managerial function enhancement is an essential culture of kaizen costing before removing nonvalue adding activities in the cost reduction phase. Kaizen costing entails managing value with cost as the focus. The kaizen costing execution stage depends on the work breakdown structure and cost limits established during cost planning. As indicated in Figure 3.1, cost/benefit appraisal is conducted continually to evaluate and identify cost gaps. Cost gaps are alternatives identified during the plan-do-check-act principle under an established standardised process.

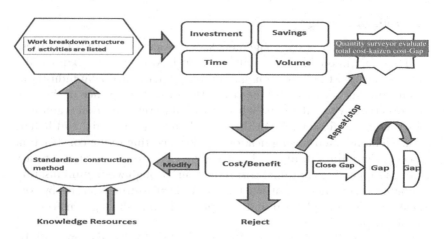

Figure 3.1 Kaizen costing process.

Source: Authors.

Kaizen costing was combined with target costing by Robert and Granja (2006) to demonstrate how cost targets can be executed during the planning stages of a construction project with the plan-do-check-act principle. Omotayo *et al.* (2020) opined that construction companies' organisational strategy and accounting system can incorporate kaizen costing for efficient adoption in the planning and construction phase. Kaizen costing can be a useful cost controlling method when combined with traditional cost controlling techniques and target costing.

3.12 Summary

Construction cost management methods are drawn from the cost management systems. Cost estimation is taken out of traditional cost management. Value planning, engineering and analysis are subsets of the value management system. Kaizen costing is a method within a continuous improvement system. However, there are also other cost management methods such as whole life cycle costing, life cycle costing, target value design, and costing that stand alone and are not associated with any system.

The next chapter continues this discussion under construction cost management techniques.

References

Alves, T. da C. L., Lichtig, W. and Rybkowski, Z. K. (2017). Implementing target value design: Tools and techniques to manage the process. *Health Environments Research and Design Journal*, 10(3), pp. 18–29. doi: 10.1177/1937586717690865

Cambridge Dictionary. (2021). Available from: https://dictionary.cambridge.org/dictionary/english/cost (Accessed 8, February 2021).

Cariaga, I., El-Diraby, T. and Osman, H. (2007). Integrating value analysis and quality function deployment for evaluating design alternatives. *Journal of Construction Engineering and Management*, 133(10), pp. 761–770. doi: 10.1061/(asce)0733-9364

CBPP. (1999). Survey of construction clients' satisfaction: Headline results, A Joint Construction Clients' Forum, Construction Best Practice Programme and Carmague report.

Cheah, C. Y. J. and Ting, S. K. (2005). Appraisal of value engineering in construction in Southeast Asia. *International Journal of Project Management*, 23(2), pp. 151–158. doi.org/10.1016/j.ijproman.2004.07.008

Cole, R. J. and Sterner, E. (2000). Reconciling theory and practice of life-cycle costing. *Building Research & Information*, 28(5–6), pp. 368–375. doi: 10.1080/096132100418519

El-Haram, M. A., Marenjak, S. and Horner, M. W. (2002). Development of a generic framework for collecting whole life cost data for the building industry. *Journal of Quality in Maintenance Engineering*, 8(2), pp. 144–151. doi: 10.1108/13552510210430017

Free Dictionary. (2021). Available from: https://encyclopedia2.thefreedictionary.com/methods (Accessed 8, Februrary 2020).

Gayani, K., Uthpala, R. and Dushan, S. (2016). Integrating sustainability concepts and value planning for sustainable construction. *Built Environment Project and Asset Management*, 6(2), pp. 125–138. doi: 10.1108/BEPAM-09-2014-0047

Kishk, M., Al-Hajj, A., Pollock, R., Aouad, G., Bakis, N. and Sun, M. (2003). Whole life costing in construction: A state of the art review. *RICS Research Paper Series*, 4(18), pp. 1–66. http://hdl.handle.net/10059/1085

Leu, S. S. and Lin, Y. C. (2008). Project performance evaluation based on statistical process control techniques. *Journal of Construction Engineering and Management*, 134(10), pp. 813–819.

Lin, B., Collins, J. and Su, R. K. (2001). Supply chain costing: An activity-based perspective. *International Journal of Physical Distribution & Logistics Management*, 31(10), pp. 702–713. doi.org/10.1108/EUM000000000628

Omotayo, T., Awuzie, B., Egbelakin, T. and Obi, L. (2020). AHP-systems thinking analyses for kaizen costing. *Buildings*, 10(12), p. 230.

Omotayo, T., and Kulatunga, U. (2015). The need for kaizen costing in indigenous Nigerian construction firms. In *International Postgraduate Research Conference*, Media City, Manchester, University of Salford, UK. June, 2015.

Pishdad-Bozorgi, P. and Gao, X. (2018). Planning and developing facility management-enabled building information model (FM-enabled BIM). *Automation in Construction*, 87, pp. 22–38. https://doi.org/10.1016/j.autcon.2017.12.004

Robert, G. and Granja, A. (2006). Target and kaizen costing implementation in construction. In *Proceedings of the 14th Annual Conference of the International Group for Lean Construction, IGLC-14 Understanding and Managing the Construction Process: Theory and Practice*, Santiago, pp. 91–105.

RICS (2012) [Online]. RICS NRM: New Rules of Measurement. Available from: https://www.rics.org/uk/upholding-professional-standards/sector-standards/construction/rics-nrm-new-rules-of-measurement/ (Accessed 4, March 2021).

Shen, Q. and Liu, G. (2003). Critical success factors for value management studies in construction. *Journal of Construction Engineering and Management*, 129(5), pp. 485–491. doi: 10.1061/(asce)0733-9364

The Merriam-Webster Dictionary (2021). Methods. Available from: https://www.merriam-webster.com/dictionary/methods (Accessed 5, March 2021).

Virle, R. and Mhaske, S. (2013). Application of earned value and earned schedule to construction project. *International Journal of Scientific Engineering and Research (IJSER)*, 1(1), pp. 1–6.

Vivan, A. L., Ortiz, F. A. and Paliari, J. (2015). Model for kaizen project development for the construction industry. *Gestão & Produção*, 23, pp. 333–349.

Zimina, D., Ballard, G. and Pasquire, C. (2012). Target value design: Using collaboration and a lean approach to reduce construction cost. *Construction Management and Economics*, 30(5), pp. 383–398. doi: 10.1080/01446193.2012.676658

4 Construction Cost Management Techniques

Temitope Omotayo, Udayangani Kulatunga, and Bankole Awuzie

4.1 Introduction: Construction cost management techniques

Techniques as a nomenclature in construction cost management are defined as systematic procedures, outlines, or routines for carrying out a task (The Business Dictionary, 2021). The Free Dictionary (2021) and Oxford Dictionary (2021) describe technique as a skill set required to accomplish an objective. Hence, techniques in construction cost management depend on some skill set acquired through training or experience. Techniques deliver the final activities required to complete tasks established in methods and systems. The quantity surveyor or cost consultant conducts cost management activities such as cash flow forecasting, interim valuation of work completed on-site, monitoring, evaluation, and reapportioning of cost-related activities on construction sites. Construction cost management techniques depend on the output of construction cost management systems and methods. For instance, interim valuations are based on cost estimation and the bills of quantities (BOQs), which are both products of the traditional cost management system. The following sections will explain the workings of key construction cost management techniques as applicable in the construction industry.

4.2 Interim valuation

In the post-contract cost control stage after the award of contract, interim valuations are conducted to evaluate the cost performance of construction projects for the purpose of payment and forecasting. In this construction phase, cost monitoring and control are indispensable. Thus, an interim valuation is derived from the physical on-site measurement of completed construction work to ascertain the cost incurred on-site. Ashworth and Perera (2015) described interim valuations as on-site accounting activities leading to the issuance of an interim certificate of payment. Interim valuations are carried out by the quantity surveyor to verify contractor's request for payment and claims. Interim valuation is a common post-contract cost

DOI: 10.1201/9781003176077-5

Table 4.1 Features of interim valuation

Cost management technique	Components	Tools	Strengths	Weaknesses
Interim Valuation	Cost information Interim certificates Claims	Working budget (bill of quantities) Final cost plan Invoices or expenses On-site measurement of work completed Digital payment platforms MS Excel Spreadsheet	Interim certificates issued during payment Permits the client to make payment based on work completed on-site	Claims and delays on the site could delay project activities Client satisfaction of work completed at every stage will inform decisions for payment certificate

controlling technique that is also tied to the traditional cost management system.

Interim valuations are based on the working budget produced as bills of quantities (BOQs) and the final cost plan. The process of creating interim valuations makes use of spreadsheets or other similar digital tools. There is a monthly or milestone comparison of planned cost and activities in the working budget and the actual cost of work completed on-site. Interim valuations are closely associated with the cash flow registers.

The main tools used to achieve an interim valuation are on-site measurements of completed work on-site (see Table 4.1). In recent times, digital tools such as CCTV cameras, drones, and virtual reality have also been used to monitor activities on construction sites. Further research into the application of remote on-site valuation of construction work may reduce the physical measurement of works completed on construction sites. The post-contract cost controlling technique of interim valuations makes it easy for clients to pay contractors based on their performance in delivering construction projects according to milestones. Conversely, interim valuation has been criticised for the occurrence of claims and delays in issuing interim certification. Nevertheless, there is a high level of accuracy of cost information produced through the interim valuation process which is unmatched by other post-contract cost controlling techniques.

4.3 Cash flow forecasting

Cash flow forecasting is a form of probabilistic costing technique embedded in the post-contract cost controlling stage of a traditional cost management system and other similar systems such as BIM, and the value management system. Cash flow forecasting is ubiquitous and a common post-contract cost controlling technique in the construction industry. Cash flow forecasting considers the unit cost of expenses incurred against the cash inflow from interim valuations as parameters for calculating contractor's profit margin and the direction of the cost (Cheng, Cao & Herianto, 2020; Van den Boomen, Bakker, Schraven & Hertogh, 2020). Cash flow registers in the form of a spreadsheet are the main tools used to determine cost outputs indicating the profit margin and final construction cost projections. Monthly financial statements are also produced as reports of cost and expenses of works carried out on construction sites.

Cash flow forecasting has been explored for its accuracy in predicting the final construction cost using deep learning (Cheng et al., 2020). Cash flow forecasts may be inaccurate in the following instances: if the quality of the cost data collected is low; when the construction cost expert is not skilled in cash flow calculations; and owing to other external economic factors, such as inflation, interest rates, exchange rate and prices of construction materials.

Cash flow forecasting is used by the employer's quantity surveyor for monitoring and managing variations, cost overruns, monthly statements, and the overall progress of the project. The contractor makes used of cost forecasting for monitoring the profit margin and predicting the final cost and profit. The S-curve is produced once there is a cost-value reconciliation. The S-curve illustrates the direction of the final construction cost against the budgeted curve. As shown in Table 4.2, some of the tools used in producing cost forecasts are the registers containing invoices of all forms of expenses. Claims letters are deployed when there is a variation in construction or a requirement for any form of additional payment. Cash flow forecasting is commonly used by contractors on all construction projects and may be applied under all construction cost management systems and methods.

Table 4.2 Features of the cash flow forecasting

Cost management technique	*Components*	*Tools*	*Strengths*	*Weaknesses*
Cash flow forecasting	Variation management Monthly financial statements Cost-value reconciliation S-curve Contractor's profit margin Manage cost overrun	Cash flow register Invoices of expense Claim letter MS Excel Spreadsheet Artificial intelligence	Allows the contractor to calculate the profit and other expenditure Permits the employer to management variations and cost overrun	Knowledge and skills of forecasting essential Output of forecasting may be inaccurate

4.4 Monitoring labour, material, equipment, and overheads

The monitoring of labour, material, and equipment cost is carried out through a thorough review of expenditure and management of overheads. Construction overheads are costs emanating from administrative charges and all fixed costs from the preliminary items of works. Omotayo, Awuzie, Egbelakin, Obi, and Ogunnusi (2020a) explained that overhead cost in construction cannot be eliminated in the construction process but can be kept within a benchmark for a suitable contractor's profit margin. Material cost fluctuations from importations, deliveries, and exchange rate can limit the extent to which cost can be effectively monitored. Hence, cost plans and the BOQs must create an allowance in terms of percentage for changes in material prices, interest rate, and exchange rate.

Ashworth and Perera (2015) noted that a well-developed risk register can also be used to document cost implications of identified risks. When the risks occur during construction, the allocated cost to the risk may be taken from the contingency fund. Another cost-planning tool is a schedule of materials which is an important tool for monitoring the cost of construction materials purchased. Depreciation cost is also monitored and calculated as part of the cost plan (Table 4.3).

It is evident that construction cost control is impossible without monitoring. Furthermore, the term "construction cost control" is only made possible by monitoring all activities on-site and feeding the output of construction

Table 4.3 Constituents of monitoring-related techniques

Cost management technique	Components	Tools	Strengths	Weaknesses
Monitoring labour, material, equipment, and overheads	Error correction Review of cost related to preliminary items of work Monthly and final financial statement	Bills of quantities Contingency fund Schedule of materials Monitoring of all cost on-site Invoices from expenditure and payments Depreciation cost Web-based payment platforms Artificial intelligence and blockchain technology for payment MS Spreadsheet and Word	A common efficient and effective cost-controlling technique	Personal heuristics and experience are required to ensure effective monitoring of cost Modern digital tools require constant training and change management

cost monitoring endeavours leads to cash flow registers and forecasts, error corrections, interim valuations, and monthly and final financial statements. All post-contract cost-controlling activities are predicated on the outcome of monitoring labour, material, equipment, and overhead cost. Construction cost-monitoring undertakings have been premised on the personal heuristics and experience of the professional performing the activities. However, this is changing considering the increasing utilisation of emerging technologies such as artificial intelligence and blockchain technology in predicting and monitoring construction costs in recent times. Moreover, regular training sessions are required on modern digital tools, which are applicable in construction cost monitoring. Monitoring of construction costs remains central to the performance of all construction cost management systems, methods, and techniques.

4.5 Cost-benefit ratio

Cost-benefit ratio is a technique under cost-benefit analysis. Cost-benefit analysis is an evaluation technique used before and during construction for the purpose of feasibility, viability appraisal, and forecasting. The cost-benefit ratio technique is used to appraise investments in a project. In the construction process, the cost-benefit ratio is associated with the value engineering construction cost method. The technique ascribes between 0 and 1 for construction materials, labour, equipment, and overhead. If the values are below 0, then the decision to supply or procure construction material or equipment may be substituted for a better alternative.

The cost-benefit ratio technique is rarely used during construction because it may lead to confusion when the project's cost deviates from what has been prepared in the working budget (see Table 4.4). The

Table 4.4 Features of cost-benefit ratio

Cost management technique	Components	Tools	Strengths	Weaknesses
Cost-benefit ratio	Projected final construction cost	Value engineering Database of alternative materials, suppliers and equipment Cashflow register	Can be combined with value engineering Usually combined with monitoring of overheads Informs the cash flow projections	Final costs and calculations may be inaccurate It may be difficult to evaluate value from cost and benefit during construction It may affect available working budget and thus, lead to variations

construction cost professional requires an accurate database of alternatives, access to the cash flow register, and continual consultations with suppliers and subcontractors for informed calculations and projections. The outcome of cost-benefit ratio analysis in construction cost evaluation may be subject to external economic factors such as inflation, fluctuations in exchange and interest rates, and the prices of construction materials. This cost-controlling technique is a viable option for investment appraisal in the pre-tender phase of construction.

4.6 Incremental milestone

Another construction cost controlling technique that has been portrayed as effective is the incremental milestone. Incremental milestones work effectively with the earned value analysis construction management method. They compute the cost of construction by examining the actual cost and actual work against the planned cost and activities on-site. Incremental milestones are also used to measure completed tasks and outline the cost for further calculations. The cost of sub-tasks within a work breakdown structure is calculated individually rather than as a bulk cost.

Table 4.5 Constituents of incremental milestone

Cost management technique	*Components*	*Tools*	*Strengths*	*Weaknesses*
Incremental milestone	Breakdown of works completed in detail Variance tracking	Programme of works Earned value analysis Invoices Cash flow register Activity-based costing MS Excel Spreadsheet	This technique identifies the work which has to be done and is useful for performance measurement	More effective in shorter durations Only effective when there is no detailed working budget

With the aid of a programme of works such as a Gantt chart, the start and finishing point of the subtasks are indicated. Cash flow registers and invoices are useful tools for incremental milestone calculation (see Table 4.5). Incremental milestones are only useful when there is no working budget or detailed cost plan and the progress of the construction project is difficult to track. Incremental milestones can be used as a technique for traditional construction cost management and adopting the cash flow register for forecasting. They depend on cost data extracted from invoices.

4.7 Identifying cost overruns

The approach of consciously identifying construction cost overruns is a major technique that has been ignored by many researchers in construction cost management and by practitioners. Omotayo, Awuzie, and Olanipekun

Table 4.6 Features of identifying cost overrun technique

Cost management technique	Components	Tools	Strengths	Weaknesses
Identifying cost overruns	Cost forecasting reports Monthly and final construction account/ statements Site meetings	Risk register Construction cost monitoring of all materials, labour, equipment, and overheads Cash flow register Earned value analysis Cost negotiation with suppliers and subcontractors Value engineering Interim valuations	Applicable in construction monitoring of project cost Cost overrun is detected early	Focussing on cost instead of value may reduce the quality of workmanship and built asset

(2020b) assessed the ranking of cost overrun identification within a list of post-contract cost controlling techniques and found that it is one of the least applied construction cost management techniques. Considering the enormity of the implications of cost overruns in construction projects, there is a need to monitor activity cash flow registers and forecasts for deviations in costs. Therefore, all construction cost management systems, methods, and techniques are important in identifying construction cost overruns. More significantly, the monthly site meeting and statements form a major input in ensuring that construction cost does not spiral out of control.

Focussing on the causations of construction cost management may devalue the output of the construction process and the final product (see Table 4.6). Additional design, cost, and management planning may lead to rework, variations, claims, and disputes, which may result in construction cost overrun. Hence, it is imperative to focus on planning construction projects using new digital tools and BIM. Cost overrun can be detected early with an adequate risk register and a review process. Site meetings are also important for investigating plausible future causations of cost overrun. The overall aim of construction cost management systems, methods, and techniques is to deliver construction projects within the budgeted cost and duration. Identifying cost overruns in the execution phase of a project can ensure the successful attainment of the aforesaid project aim.

4.8 Managing variations

Variation can be described as all construction activities outside the scope of the design and cost requiring changes, rework, or error correction. In every construction project, there may be minor or major

Table 4.7 Features of variation management

Cost management technique	Components	Tools	Strengths	Weaknesses
Managing variation	Variation orders Rework Site meetings	Site supervision and reports Monitoring of all on-site activities including construction cost Cash flow registers Incremental milestones Earned value analysis Interim valuations	Essential in some cases where the project has experienced sudden change	If not properly implemented, cost and time overruns may occur

variations. Variations that do not require major changes such as demotions and rework or demand major cost are considered to be minor variations (Alsuliman, Bowles & Chen, 2012). Major variations occur when the changes necessitate urgent rework, demolition, and major cost. Hence, the monitoring of all on-site activities, including construction cost, cash flow registers, incremental milestones, earned value analysis, and interim valuations, is necessary for ensuring that major and minor variations are managed.

Table 4.7 summarised the weaknesses, strengths, tools, and components of how to manage variations. The weakness of managing variation is simply the consequence of not managing variation orders effectually. Managing construction variations is an important technique for ensuring construction projects are delivered within budget and planned schedule.

4.9 Site meetings and post-project reviews

Intermittent construction-site meetings can ensure proper cost management. Site meetings may be monthly or weekly and can aid project documentation and construction cost monitoring and management. The outcomes of project site meetings can lead to improved post-project reviews. Site meetings and post-project reviews act as a feedback mechanism for current and future construction projects (Omotayo, Boateng, Osobajo, Oke & Obi, 2019). The purpose of site meetings and post-project reviews goes beyond managing construction cost; it is about creating a knowledge database involving graphical and non-graphical data (Carrillo, Harding & Choudhary, 2011). Performance evaluation in earned value analysis, value engineering and analysis, cash flow forecasts, and incremental costs depends on the outcomes of reports produced during site meetings and post-project reviews. In terms of construction cost management,

Table 4.8 Features of site meeting and post-project review

Cost management technique	Components	Tools	Strengths	Weaknesses
Site meetings and post-project reviews	Project reports and evaluation Knowledge database for historical cost	All cost controlling technique tools All digital tools such as MS Project, MS Excel Spreadsheet and Word Power BI	An important platform for handling existing and future projects cost Ensures forgoing errors are not transferred to new projects Provides a platform for continuous cost improvement and management	Constant meeting and reviews may be daunting tasks

expenditure and profits are evaluated in site meetings for continuous cost improvement purposes.

The benefits of site meetings and post-project reviews, as indicated in Table 4.8, provide a useful platform for managing existing and future project costs. Errors are corrected when a joint review of the project costs takes place. However, the process of post-project reviews may be difficult to conduct, especially in very large projects. In this regard, artificial intelligence tools such as Power BI may be used to manage and analyse large and complex construction cost data for documentation and continuous improvement drives.

4.10 Historical cost data and profit and loss summary

Historical cost data from site meetings and post-project reviews are vital for cost forecasting using cash flows, earned value analysis, and incremental milestones. Applying historical data in post-contract cost control implies that the construction project details are similar to the one to which it is being applied and that some form of cost performance measurement was conducted to ascertain the veracity of the data. The application of historical data in other construction cost-controlling techniques will be improved when further site meetings and post-project reviews are held. More significantly, the profit and loss summaries are important documents in determining the cost performance. The profit and loss account forms part of the historical cost data.

Table 4.9 indicated that the benefit of applying historical cost data in a new project is the ease of using similar data from similar designs as part of an expert judgment. If there are unidentified errors in these historical cost

Table 4.9 Constituents of the application of historical data and profit and loss summary

Cost management technique	Components	Tools	Strengths	Weaknesses
Historical cost data and profit and loss summary	Organisational construction cost database Final account statement	Cash flow register and forecasts Incremental milestone Earned value analysis Value analysis Value engineering Construction cost database such as BCIS	Ensures important cost data are utilised by construction project team during construction An important technique in expert judgement	Usually influenced by market forces and rebasing is required The final cost output may not be accurate because of unidentified errors in previous projects

data, such unidentified errors will affect overall cost performance; hence, a loss. Efforts in rebasing historical cost data can improve the final outcome of the profit and loss account summary. Different construction projects have varying conditions, climates, topographies, and ground conditions that may not permit the transfer of historical cost data to new projects.

4.11 Plan-do-check-act (PDCA) cost reduction principle

Kaizen costing is a method of continuous improvement in construction cost management. Vivan, Ortiz, and Paliari (2015) observed the application of kaizen in post-contract construction cost management. Findings from the study by Vivan et al. (2015) reported the application of the plan-do-check-act (PDCA) principle in reducing construction cost. Omotayo et al. (2020a) concluded that continual reduction of overhead cost and inclusion of post-project reviews in post-contract cost control were key factors to attaining continuous improvement through the PDCA principle. PDCA can be used in cost planning and control.

PDCA in construction cost management leads to improved profitability, waste minimisation, better quality, and client satisfaction (see Table 4.10). The drawback of the PDCA approach in construction cost management is that it is still new to the knowledge domain of construction cost and project management professionals. Regular training will be required to update their knowledge and skills in applying PDCA in construction cost management. Also, the PDCA approach may delay the progress of a construction project where inexperienced construction professionals are still trying to identify cost-saving measures. PDCA in cost management can

Table 4.10 Components of the plan-do-check-act in construction cost management

Cost management technique	Components	Tools	Strengths	Weaknesses
Plan-do-check-act cost reduction principle	PDCA register	Kaizen costing Site planning and management Historical cost data Cashflow register Site meetings and post-project reviews Interim valuation	Important process in achieving continuous improvement in construction cost Waste reduction, improved profitability, better quality, and client satisfaction	Training is required to apply PDCA in construction cost management The PDCA process may lead to delays in construction project, if it is not managed by a kaizen costing expert

be successful with the inclusion of historical cost data, site meetings and post-project reviews, cash flow registers, interim valuation, and proper site management and planning.

4.12 Post-contract cost control

Post-contract cost is the aggregation of costs associated with the construction phase after the award of a contract. Immediately after cost estimation activities, tendering, and contractor selection, it is necessary to control construction cost. The aim of post- contract cost control is to control the budgeted resources during construction activities. The resources used in construction projects are mainly construction materials, plant and equipment, labour, information technology, and professional and skilled workers. Construction resources are limited and must be maximised for the effective delivery of construction projects. Hence, the quantity surveyor has the responsibility to manage the limited construction resources. The techniques employed in managing the limited construction resources are cash flow registers, cost value reconciliation, and earned value analysis. The contractor will have to monitor expenses and financial inflow for the purpose of profitability, while the client wants to ensure there is no overspending which may lead to cost overruns. Post-contract cost control is based on mitigating the risks which may cause cost overrun. Thus, from the perspectives of both the contractor and the client, construction cost must be managed effectively.

Figure 4.1 summarises all the links between the construction cost management systems, methods, and techniques. The systems are orange, the methods are blue, and the greyed sections contain the techniques.

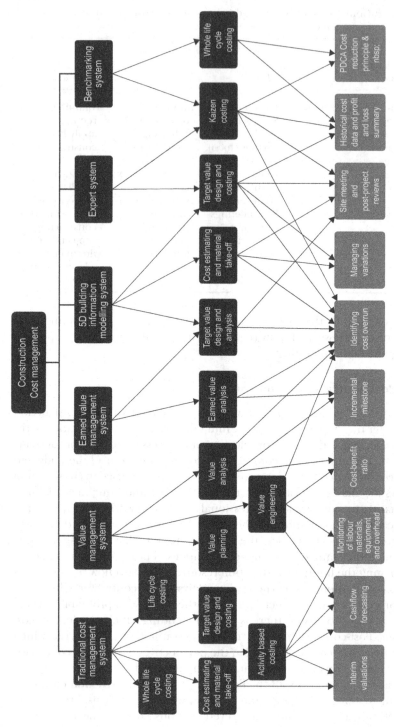

Figure 4.1 Construction cost management systems, methods, and techniques.

Source: Authors.

Continuous improvement in construction cost management can be achieved by not only applying the PDCA in cost management but also by embedding PDCA within every other technique. Therefore, traditional cost management systems can incorporate the PDCA principle in cost estimation and post-contract cost controlling techniques.

References

Alsuliman, J., Bowles, G. and Chen, Z. (2012). Current practice of variation order management in the Saudi construction industry. In *Proceedings of the 28th Annual Conference, Association of Researchers in Construction Management (ARCOM)*, 2 September, Edinburgh, UK, pp. 1003–1012.

Ashworth, A. and Perera, S. (2015). *Cost studies of buildings (6th edition)*. Routledge: London.

Carrillo, P., Harding, J. and Choudhary, A. (2011). Knowledge discovery from post-project reviews. *Construction Management and Economics*, 29(7), pp. 713–723. doi: 10.1080/01446193.2011.588953.

Cheng, M. Y., Cao, M. T. and Herianto, J. G. (2020). Symbiotic organisms search-optimized deep learning technique for mapping construction cash flow considering complexity of project. *Chaos, Solitons & Fractals*, 138, p. 109869. doi: https://doi.org/10.1016/j.chaos.2020.109869.

Free Dictionary. (2021). Available from: https://encyclopedia2.thefreedictionary.com/methods (Accessed 8, February 2020).

Omotayo, T. S., Boateng, P., Osobajo, O., Oke, A. and Obi, L. I. (2019). Systems thinking and CMM for continuous improvement in the construction industry. *International Journal of Productivity and Performance Management*, 69(2), pp. 271–296. doi: 10.1108/IJPPM-11-2018-0417.

Omotayo, T., Awuzie, B., Egbelakin, T., Obi, L. and Ogunnusi, M. (2020a). AHP-systems thinking analyses for kaizen costing. *Buildings*, 10(12), p. 230.

Omotayo, T., Awuzie, B. and Olanipekun, A. O. (2020b). An artificial neural network approach to predicting most applicable post-contract cost-controlling techniques in construction projects. *Applied Sciences*, 10(15), pp. 5171. doi: 10.3390/app10155171.

Oxford Dictionary. (2021). Available from https://www.oxfordlearnersdictionaries.com/definition/english/price_1?q=price (Accessed 8, February 2021).

The Business Dictionary. (2021). System [online]. Available from: https://www.dictionary.com/browse/system (Accessed 3, February 2021).

Van den Boomen, M., Bakker, H. L. M., Schraven, D. F. J. and Hertogh, M. J. C. M. (2020). Probabilistic life cycle cash flow forecasting with price uncertainty following a geometric Brownian motion. *Structure and Infrastructure Engineering*, AHEAD-OF-PRINT, pp. 1–15. doi: 10.1080/15732479.2020.1832540.

Vivan, A. L., Ortiz, F. A. and Paliari, J. (2015). Model for kaizen project development for the construction industry. *Gestão & Produção*, 23, pp. 333–349.

5 Modern Methods of Cost Control

Temitope Omotayo, Udayangani Kulatunga, and Bankole Awuzie

5.1 Modern methods of construction and cost control

Recent developments in construction and computing have given rise to new methods of controlling construction cost. These developments emanate from artificial intelligence (AI), 3D-printed homes, blockchain technology for payments in construction, Intranet-based costing, mobile technology in construction projects, and generally internet-of-things in construction project management. Sustainable construction in the form of prefabricated construction, ecohomes, passive houses, and buildings with high-energy ratings have their own distinct costs. This chapter discusses the cost control in emerging construction technologies and how they are controlled during the execution phase. This chapter provides an overview of cost components and cost control in various modern methods of construction. The basis of modern methods of construction is sustainability. Green costing is a term in sustainable construction that considers the typology of construction materials, methods of evaluation, energy usage in buildings, big data applications, and reduction of carbon emission in the construction process and in-use phase (Illankoon & Lu, 2019; Orji & Wei, 2016). The cost of constructing new built asset requires new control measures. These modern methods of construction are in relation to each type of sustainable construction method.

5.2 The cost of prefabricated construction

Prefabrication is a process whereby building construction elements are manufactured in the factory before being transported to the site (Hong, Shen, Li, Zhang, & Zhang, 2018; Robichaud & Anantatmula, 2011). In their study of eight building construction projects in China, Hong et al. (2018) observed that costs associated with concrete and steel accounted for 26% to 60% of the total construction cost. Also, they determined that this was closely followed by labour costs which ranged from 17% to 30% of the total construction cost. A comparison of costs associated with traditional methods of construction with prefabricated construction indicates

DOI: 10.1201/9781003176077-6

a higher cost in prefabrication in China. In the United Kingdom (UK), Pan and Sidwell (2011) studied cross-wall materials (cross-wall construction is a type of wall that bears the load for the precast floor above it) in 20 medium high-rise buildings. Pan and Sidwell (2011) observed that cross-walls were significantly 11% to 32% cheaper than reinforced concrete and steel. This lower cost in prefabricated buildings will provide a cheaper alternative to the conventional construction method.

Over the years, the potential of innovative building material such as hardwood cross-laminated timber, self-healing concrete, bioplastic, home-ostat façade, artificial spider silk, 3D-printed graphene, and aerographite to drive down the cost of prefabricated construction as opposed to the use of conventional materials such as concrete, steel, and block has been noted Fang and Ng (2011). Furthermore, a comparison of sustainable prefabricated buildings and traditional site-built buildings was conducted by Dobson Sourani, Sertyesilisik, and Tunstall (2013). In their assessment of the component and material supply chain in prefabricated buildings against that of traditional buildings, Dobson et al. (2013) established that a long-term supply chain of building materials can be created in prefabricated buildings. An integrated supply chain is one of the advantages of prefabricated construction over site-built projects because it reduces overhead costs associated with production and unstructured delivery of construction materials. Table 5.1 explains the major cost benefits of prevaricated construction over the conventional construction method.

Compared with on-site buildings, prefabricated buildings provide higher quality at a lower cost and time. Higher quality is achieved through the adoption of a standardised process of producing building elements in the factory. This process makes it possible for cost reduction with the aid of lean or agile manufacturing concepts. The supply of building materials used in off-site fabrication of building elements can lead to an integrated supply chain. In on-site construction, the supply chain has been criticised for not being integrated in the construction life cycle and stakeholder management (Magill, Jafarifar, Watson, & Omotayo, 2020). In terms of the schedule and cost of construction, the prefabrication process makes it easier for cost reduction as incidents of cost overrun are managed effectively. Cost overrun is also mitigated by reduced occurrences of change orders. Change orders emanate from variations in design, interpretation, errors, or any form of inconsistencies impacting the delivery of construction processes. The location of construction plant and equipment in off-site construction contributes to cost reduction because plant and equipment transportation and hire costs are eliminated. Finally, higher quality of properties delivered through off-site construction usually leads to early defect detection and rectification as well as a lower maintenance cost when compared to on-site constructed properties.

The only challenge with prefabricated properties is the initial cost of setting up the factory and design for manufacturing approaches which may

Table 5.1 Cost factors in prefabrication and site-built projects

Cost factors	Prefabrication	Site-constructed
Quality	Higher quality can be achieved at a lower cost and duration	Depends on the quality of workmanship, site conditions, and cost
Building element material supply	Integrated supply chain can be established	Supply of materials is based on each project
Schedule length and reliability	Construction time and cost can be reduced with a longer lead time. Incidents of cost overrun are reduced	Longer construction delay arising from shorter lead time. Incidents of cost overrun are prevalent
Coordination time	Needs more coordination between factory and site construction	Additional time is still required for adjustment of dimension and changes
Flexibility	Not flexible to changes when assembling on construction sites	Flexible but in terms of limited changes on construction sites
Impact of change orders	Becomes more expensive, cost and time overrun, difficult to change, especially in large-scale projects	It can better accommodate changes, but cost and time overrun may occur
Delivery and shipping	The location of the plant determines delivery and shipping cost	Raw material delivery requires shipping cost
Maintenance cost	Higher quality will lead to the minimal cost of operation and maintenance	Operations and maintenance cost usually rises from the location and quality of construction

Sources: Adapted from Pan and Sidwell (2011); Robichaud and Anantatmula (2011).

vary in different countries. Construction cost control is easier in standardised construction processes and new concepts in the construction cost improvement to be implemented. The integrated supply chain, coordination time, flexibility, impact of change orders, and maintenance costs are reduced with the application of prefabricated construction. Continuous cost improvement is therefore more probable to implement in a factory setting.

5.3 Cost of 3D-printed buildings

3D-printed buildings are specially constructed buildings that are produced by the use of super-size printers to produce layers of a special concrete that culminates in the building envelope and its internal components (Wu, Wang, & Wang, 2016). 3D printing in the construction industry is a form of additive manufacturing and its application in delivering sustainability remains relatively unexplored. The first 3D printer was invented by Charles Hull in 1983 (Sakin & Kiroglu, 2017). The modern application of 3D printing is still very recent in the construction industry. However, evidence from such projects has identified a reduction in labour cost by up to

Table 5.2 Elemental cost plan for Apis Cor 3D-printed house

S/N	Element	Cost/m^2	Amount (£)
1	Foundation	5.27	200.24
2	Concrete walls	30.89	1173.96
3	Floor and roof finishing	46.30	1759.50
4	Electrical installation	4.60	174.94
5	Windows and doors	67.49	2564.79
6	Finishes to the external wall	15.81	600.72
7	Finishes to the internal walls and ceiling	22.41	851.56
	Total cost	192.77	7,325.71

Source: Adapted from All3DP (2021).

80% and construction waste between 30% and 60% as benefits accruable from its uptake (Sakin & Kiroglu, 2017). Other examples of buildings that were 3D printed at lower cost abound in the literature.

Dubai's Office of the Future was 3D printed to provide an office space for the United Arab Emirates National Committee headquarters for the Dubai Futures Foundation (Sakin & Kiroglu, 2017). In constructing the Dubai Office of the Future, labour cost was reduced by 50%. The reduction or elimination of cement usage in the concrete mix along with the duration of printing are two major cost components that facilitate cost reduction (Du Plessis, Babafemi, Paul, Panda, Tran, & Broeckhoven, 2021; Zhang et al., 2019).

Another example where 3D printing led to a reduction in construction cost is the construction of the 38 m² Apis Cor in Russia. The elemental cost breakdown for the Apis Cor is shown in Table 5.2.

From the elemental cost breakdown of the Apis Cor 3D-printed building, it can be inferred that the cost of equipment and plants required for building construction can be eliminated. Higher costs in 3D-printed buildings depend on the size. The larger the size, the higher the cost. Notwithstanding, the overall cost of construction when compared with traditionally site-constructed building can be reduced by 60% with 3D printing. The cost of a large concrete 3D printer is about £7,300 and the cost is gradually reducing owing to the demand (All3DP, 2021). The main requirement for executing 3D-printed housing is the 3D printer and the technical know-how in operating the 3D printer. Furthermore, the inclusion of 3D modelling and BIM may be essential in facilitating the delivery of 3D-printed buildings (Du Plessis et al., 2021; Sakin & Kiroglu, 2017). In the United States (US), the iconic Apis Cor building was constructed in under 24 hours at a cost of £7,200; China's Winsun of 66 m² was also built within 24 hours at a cost of £3,500 (All3DP, 2021). The application of 3D-printed houses depends on the location of construction while other cost factors such as land acquisition and approval cost may vary.

The concept of continuous cost improvement is based on the standardisation of the construction process, waste elimination, and overhead cost reduction. 3D-printed building is an avenue to adopt continuous cost improvement practically through the elimination of physical and non-physical waste and overhead costs. Hence, it can be deduced that there is a subconscious application of continuous cost improvement in the construction industry through 3D printing. Nonetheless, the lack of awareness, technical know-how and the cost of the 3D printer are all major barriers limiting the wider adoption of 3D printing in the construction industry.

5.4 Artificial intelligence in construction cost control

The construction industry is still far behind other economic sectors as it concerns AI application (Blanco, Fuchs, Parsons, & Ribeirinho, 2018). AI is a machine learning algorithm capable of recognising patterns, classifying themes, understanding behavioural patterns and a host of other applications (Blanco et al., 2018; Cheng, Tsai, & Liu, 2009). AI may be developed under supervised or unsupervised learning of a data set. In the construction industry, the application of AI is effective in design and cost prediction. Conversely, there have been some advances in the application of AI in cost estimation (Elmousalami, 2020). Also, Cheng et al. (2009) discussed the application of AI in predicting accurate contractors' cashflow. Optimal cashflow forecasting, as outlined in Section 4.2, results in effective construction cost management. By predicting the cost of construction accurately, the emergence or impact of cost overruns can either be avoided or mitigated. Cheng et al. (2009) made use of genetic algorithm (GA), fuzzy logic, artificial neural network (ANN), K-mean clustering, and evolutionary fuzzy neural inference model (EFNIM) to expound the idea of predicting construction cost accurately using the S-curve. The applications of AI in construction cost control begin with accurate measurement and estimation.

Wang, Yu, and Chan (2012) made use of data from 92 building projects to demonstrate how early planning can influence the final accounts of construction projects. Investigations by Wang et al. (2012) made use of ANN and support vector machines classification model to provide an accuracy of 80% project success where there was effective early cost and schedule planning.

5.4.1 Predicting behaviour of construction professionals towards construction cost control

One of the benefits of AI in the construction industry is cost prediction and behavioural pattern classification. In classifying the behaviour of construction project managers and quantity surveyors when choosing the post-contract cost controlling techniques, Omotayo, Awuzie, and Olanipekun (2020) observed that ANN can be used to predict and classify how construction project managers and quantity surveyors behave at various stages of construction. Omotayo et al. (2020) made use of small data sets and proved that small data was also relevant in ANN classification in the construction industry. During the end of the construction process, the construction project manager takes more responsibility in adopting all construction cost-monitoring activities along with historical cost. Multi-criteria decision-making through fuzzy logic analytical hierarchy process and other AI-related methods have proved useful in decision-making processes in the construction industry. Dziadosz and Kończak (2016) further noted that decision-making in the construction industry is complex and dynamic. In this sphere, AI provides an array of methods to exploit. The pattern and behaviours of decision-makers in simple or complex projects all have an overarching impact on the financial performance of construction projects. In respect of continuous improvement, the application of AI provides an opportunity to understand how construction professionals behave and how to improve their behaviour in terms of making the right decisions at the right time.

5.5 Blockchain technology and construction cost control

Blockchain technology is a major aspect of the fourth industrial revolution which the construction industry can leverage on under the auspice of continuous improvement concept in construction cost management to create more trust among the stakeholders. Blockchain technology is a ledger system of recording information in a devolved database that is difficult to hack or manipulate (Pilkington, 2016; Yaga, Mell, Roby, & Scarfone, 2019). Blockchain technology makes use of AI to advance the course of peer-to-peer payments. Blockchain technology is a decentralised system that provides encrypted and secured access to digital transactions (Turk & Klinc, 2017). There are several applications of blockchain technology in construction management. One such application is secured peer-to-peer payment through cryptocurrency. Turk and Klinc (2017) observed that BIM and blockchain technology are important concepts in facilitating continuous improvement within the construction industry. The potential application of blockchain technology in the construction industry was studied by Wang, Wu, Wang, and Shou (2017) and their assertions in relation to construction cost revealed that clients have the following:

i   An automated system of procurement and payments at lower schedule and cost.
ii Cheaper verification and notarisation of construction document, in this regard, invoices and valuation documents can be verified easily with blockchain technology.
iii More trust in payment and transactions through authentication of the construction supply chain system.
iv Smart contracts for efficient contract management.
v   Elimination of payment and cashflow issues through automated payment.

Project cost control and change management are aspects that pose many challenges to construction cost management which blockchain technology can contribute towards mitigating (Kim, Lee, & Kim, 2020). Trust issues between subcontractors and contractors emanate from late payments. In terms of improving trust between stakeholders within the construction inter-organisational setting, blockchain technology can augment all sources of trust building in the supply chain by reducing risk and cost (Qian & Papadonikolaki, 2020). Blockchain technology also improves collaboration among organisations by enabling a system and cognition-based platform for integrating the supply chain with the construction process.

Blockchain technology and cryptocurrencies can be adopted in the construction industry to reduce transaction cost by reducing overhead charges associated with payments. Additionally, merging BIM with blockchain technology can facilitate effective stakeholder management. The community is a major stakeholder in construction projects. BIM and blockchain can be used to create open-source information for communicating ideas publicly. This approach can create trust in government-led construction projects.

The potential of blockchain technology can significantly enrich the construction cost control process improvement. However, the readiness of the construction industry to change in the fourth industrial revolution era largely depends on education, attitude towards new innovations and perception of cost. The perception of cost of change and new technology has limited the adoption of innovative ideas and new technology such as blockchain in the construction industry. The merits of blockchain technology in construction cost control deals are evident in the following:

i   Timely payment with the aid of cryptocurrencies.
ii Enhancement of collaboration in the construction process by integrating it with building information modelling (BIM).
iii Trust creation in the construction process through the integration of the supply chain and other stakeholders.
iv Openness in built asset information management and sharing.

Cryptocurrency in a construction project as a blockchain payment option will ensure more trust, collaboration and cohesiveness in projects, thus eliminating disputes and unnecessary overhead costs associated with traditional cost control.

5.6 Mobile construction project management and construction cost control

Mobile construction project management is the application of smart mobile devices such as tablets and mobile phones to achieve specific project objectives such as resource monitoring, schedule recording, and expense of construction project cost (Kim, Park, Lim, & Kim, 2013; Kimoto, Endo, Iwashita, & Fujiwara, 2005). The application of mobile construction project management range from mobile payment; augmented reality (AR) for valuation of work completed on-site; reduction of rework and variation change orders; defect reduction; productivity and performance improvement (Kim et al., 2013). Mobile construction project management makes use of apps on mobile phones and tablets to manage construction information through an electronic document management (EDM) system (Al Qady & Kandil, 2010). The application of mobile phones in site monitoring of materials supply, interim valuations, and payments also interacts with the regular database where other stakeholders can view the information on their personal computers (PCs). Hence, in many instances where mobile phones and tablets are used for construction cost-controlling activities, PCs are also utilised.

Igwe, Mohamed, Dzahirand, Yusof, and Khiyon (2020) itemised the process of applying mobile construction management as the following:

i   Constant budgeting as drawn from the programme of works and contract sum.
ii Appraisal of the practicality of the proposed programme of works and budget.
iii Final approval of the programme of works and budget.
iv Monitoring of on-site activities against the proposed budget with mobile devices.
v   Cost value reconciliation of the expenses incurred on-site and the budget with a mobile application.
vi Cost information upload and storage for further application.

Mobile construction project management was tested for its effectiveness in cost control along with BIM using Android AR in a study conducted by Zaher, Greenwood, and Marzouk (2018). The findings revealed that the budgeted cost of work scheduled (BCWS) and the budgeted cost of work performed (BCWP) can be computed easily using the android BIM-enabled AR application. Zaher et al. (2018) also observed that the information process was shredded with a Microsoft Excel Spreadsheet in a cloud database such as Google Drive. AR provides a more accurate opportunity for quantity surveyors to value the work completed on-site, record the outcome, and communicate effectively. The influence of mobile project management in the construction industry can be transferred to the supply chain integration ambitions and subcontractor relationship management. A continual application of the aforementioned practice in construction can reduce construction overheads incrementally, and thus create continuous cost improvement.

Other applications of mobile construction project management are mobile BIM and lean implementation, which were crafted by Koseoglu and Nurtan-Gunes (2018). BIM and lean construction are two concepts in the process of adoption relating to many construction economies. The implementation challenges border around the cost of technology, training, integration of the supply chain, awareness, and education. Mobile BIM and lean construction combination were effective in cost saving and reduction through quality improvement and productivity enhancements (Zaher et al., 2018).

Furthermore, mobile project management makes it easier for construction professionals and workers to access various information technology (IT) applications such as Autodesk suites, Navisworks, BIM 360 Glue, MS Project, MS Excel, MS Word, and Primavera. In recent developments, most construction IT applications now have mobile versions, thereby making it easier for productivity and quality to improve continually on construction sites. When productivity improves, construction cost will improve correspondingly. The use of mobile construction project management in the construction industry largely depends on the IT skill sets of construction professionals. For instance, if quantity surveyors find it difficult to make use of regular cost-related IT applications such as MS Project, CostX, Bluebeam, and MS Excel, they will not be inclined to make use of mobile phones or tablets on construction sites.

5.6.1 Videoconferencing in construction cost control

Videoconference is a part of mobile project management and a major aspect of improvements in collaboration and communication. In the construction industry, videoconferencing sessions have become more common owing to the nature of social restrictions resulting from the COVID-19 pandemic. In exploring the applications of videoconferencing in the construction industry, Bond-Barnard, Fletcher, and Steyn (2016), studied 210 project professionals across a range of industries who provided feedback on the quality of communication in projects. Construction professionals noted that even though communication improved, the security of Internet video message and conferencing platforms posed an issue. Nevertheless, videoconferences have become important in construction projects.

Videoconferencing reduces the time and cost of travel to inspect construction sites. The application of videoconferencing does not imply that physical site inspection and monitoring will be neglected. According to the terms of many construction contracts, it is necessary to visit construction sites for the purpose of inspection, valuation, monitoring, supply of construction materials, and quality management. However, the advent of BIM and videoconferencing support construction activities, thereby leading to more collaboration and engagement in the construction industry.

The health sector has also benefitted from the use of videoconferencing for surgeries and meetings, thus creating an avenue for further adoption in the construction industry (Bower, Hinks, Wright, Hardcastle, & Cuckow,

2001). Nonetheless, construction cost monitoring and control is still an activity that demands careful and concise details from the construction sites. However, when there is a high level of collaboration in construction, the opportunity for improvement will emerge. Site meeting and post-project reviews can be conducted effectively with the aid of videoconferencing. These meetings and reviews are important when implementing continuous improvement. Cost control practices such as the monitoring of material delivery and the supply chain may also benefit significantly from videoconferencing engagements.

In situations where construction companies have multiple projects, it may become tiring and unproductive to keep travelling to every site for inspection and cost monitoring. Hence, videoconference platforms such as MS Teams, Skype, or Zoom may be used along with CCTV cameras to evaluate the progress of work on construction sites. The usage of videoconferencing in construction management may become more popular in the post-COVID-19 era.

5.7 Big data analytics and construction cost

Big data analytic solutions such as Microsoft Power BI and Zoho Analytics are becoming important in construction project management. The construction industry is information and data intensive. Hence, conventional software applications such as MS Project and Excel have not made it easier for construction professionals to evaluate the amount of data in the construction industry. Furthermore, authors such as Cheng et al. (2009), Wang et al. (2012), and Elmousalami (2020) demonstrate how big data in the construction industry can be used to predict the cost performance of construction projects.

Bilal, Oyedele, Kusimo, Owolabi, Akanbi, Ajayi, Akinade, and Davila Delgado (2019) proposed a big data architecture capable of harnessing construction data to predict the profitability of construction companies. Data from cashflows, risk identification and quantification, bid prediction and selection, cost forecasting, and other construction variables passed through a data and text mining, spatial analysis, diverse data wrangling, and integration specifications create a database for determining the profitability of construction companies. The insight created by Bilal et al. (2019) provides a machine-learning algorithm for project evaluations inclusive of cost and performance measurement of organisations.

The essence of big data analytics concurs with the tenets of continuous improvement. Considering the large datasets emanating from construction activities and processes, there is a need for a big data analysis architecture in construction companies and projects. In megaprojects such as power plants or dams, accurate estimating and construction cost control are paramount. Davila Delgado, Oyedele, Bilal, Ajayi, Akanbi, and Akinade (2020) suggested a predictive analytics and modelling system (PAMS) to evaluate a 2.75-million-point UK dataset of power transmission projects in the last 10 years. The PAMS architecture made use of BIM and predictive and prescriptive analytics, regression models, simulations, heuristics models, optimisation models, and expert systems to determine the accuracy of the cost estimates and final construction cost.

The accuracy of cost estimation and forecasting can reduce incidents of cost overruns in megaprojects when big data analytics is used to measure and predict cost performance. Continuous improvement in construction cost management depends on innovative trends such as big data analytics. However, the application of big data analytics depends on awareness and application of the outputs. Construction project cost control practices can improve when there is a better understanding of how specific variables interact with cost determinants to produce the final construction cost. In achieving this, the manipulation of big construction data and understanding of cost performance and final account of projects should not be an afterthought but an organisational policy for continuous cost improvement.

5.8 Cloud-based database and construction cost control

Construction big data analytics may be expensive for small and medium-scale construction companies to adopt. Web-based cost control makes use of a database or cloud computing. This was why Perera and Imriyas (2004) proposed the application of an integrated database combining MS Project and MS Access as an economical viable option for small and medium-scale construction companies. A web-based database management system (DBMS) can be used along with BIM to monitor construction cost. In the process of estimating cost in the planning stages, the construction rate library may be compiled in the DBMS to produce the bills of quantities. Furthermore, cost-tracking activities can be conducted and managed effectively in a DBMS (Perera & Imriyas, 2004).

The operational cost of construction companies can also be reduced with the use of an extranet system thereby increasing the profit margin (Anandarajan, Anandarajan, & Wen, 1998). The application of a DBMS can reduce constriction cost as Anandarajan et al. (1998) noted. These reductions can be evident in the following areas:

i   Data entry cost is reduced by ensuring no duplications are made in data entry. This process also reduces errors in estimation and cost control.

ii Office supplies cost can be reduced when electronic documents are used instead of paper.

iii Cost of financing accounts receivable such as payment of invoices can be reduced with quicker preparation of invoices with MS Excel, which may be stored in a cloud system.

iv Personnel cost reduction results from database storage of all documents.

v There is a reduction in communication cost by telephone and letter.

Construction project performance is evaluated effectively with the use of a web-based cloud system (Cheung, Suen, & Cheung, 2004). A project performance monitoring system (PPMS) can expedite and ease cost monitoring and performance of projects and construction companies (Cheung et al., 2004). Consequently, effective decision-making in terms of construction cost management can improve continually with the use of cloud database systems. The drawback of making use of a cloud-computing database for managing construction cost may be the inclusivity of other stakeholders such as the subcontractor and suppliers. Conversely, this challenge can easily be resolved by providing third-party rights to subcontractors and suppliers. Thus, leading to a more collaborative system of monitoring project costs.

5.9 Conclusion: Continuous cost improvement and construction new trends

Continuous cost improvement in new construction trends such as prefabrication, 3D printing, videoconferencing, AI, big data analytics, mobile project management, and cloud-based cost control present an array of opportunities for an inclusive arrangement for all construction stakeholders. These benefits have cost reduction in labour, plant, and equipment; operations cost; preliminary items of works such as office supplies, water and energy usage; and relationship management. From the studies reported in this chapter, continuous cost improvement is being practised at a very low and slow pace in the construction industry. Therefore, there is a need for more awareness, training, and encouragement of new trends originating from construction Industry 4.0.

References

All3DP. (2021). Apis Cor 3D Prints a small house in 24 hours for $10,000. Available from: https://all3dp.com/apis-cor-3d-prints-small-house/ (Accessed 22 April 2021).

Al Qady, M. and Kandil, A. (2010). Concept relation extraction from construction documents using natural language processing. Journal of Construction Engineering and Management. 136(3), pp. 294–302. doi: 10.1061/(asce)co.1943-7862.0000131

Anandarajan, M., Anandarajan, A. and Wen, H. J. (1998). Extranets: A tool for cost control in a value chain framework. Industrial Management & Data Systems. 98(3), pp. 120–128. doi: 10.1108/02635579810213125

Bilal, M., Oyedele, L. O., Kusimo, H. O., Owolabi, H. A., Akanbi, L. A., Ajayi, A. O., Akinade, O. O. and Davila Delgado, J. M. (2019). Investigating profitability performance of construction projects using big data: A project analytics approach. Journal of Building Engineering. 26(2019), pp. 1008502. doi: 10.1016/j.jobe.2019.100850

Blanco, J. L., Fuchs, S., Parsons, M. and Ribeirinho, M. J. (2018). Artificial intelligence : Construction technology's next frontier [Online]. *McKinsey & Company*. Available from https://www.mckinsey.com/business-functions/operations/our-insights/artificial-intelligence-construction-technologys-next-frontier (Accessed 6, April 2021).

Bond-Barnard, T., Fletcher, L. and Steyn, H. (2016). Exploring the influence of instant messaging and video conferencing on the quality of project communication. Acta Structilia. 23(1), pp. 36–69. doi: 10.18820/24150487/as23i1.2

Bower, D. J., Hinks, J., Wright, H., Hardcastle, C. and Cuckow, H. (2001). ICTs, videoconferencing and the construction industry: Opportunity or threat?. Construction Innovation. 1(2), pp. 129–144. doi: 10.1108/14714170110814569

Cheng, M. Y., Tsai, H. C. and Liu, C. L. (2009). Artificial intelligence approaches to achieve strategic control over project cash flows. Automation in Construction. 18(4), pp. 386–393. doi: 10.1016/j.autcon.2008.10.005

Cheung, S. O., Suen, H. C. H. and Cheung, K. K. W. (2004). PPMS: A Web-based construction project performance monitoring system. Automation in Construction. 13(3), pp. 361–376. doi: 10.1016/j.autcon.2003.12.001

Davila Delgado, J. M., Oyedele, L., Bilal, M. Ajayi, A., Akanbi, L. and Akinade, O. (2020). Big data analytics system for costing power transmission projects. Journal of Construction Engineering and Management. 146(1), 05019017. doi: 10.1061/(asce)co.1943-7862.0001745

Dobson, D. W., Sourani, A., Sertyesilisik, B. and Tunstall, A. (2013). Sustainable construction: Analysis of Its costs and benefits. American Journal of Civil Engineering and Architecture. 1(2), pp. 32–38. doi: 10.12691/ajcea-1-2-2

Du Plessis, A., Babafemi, A. J., Paul, S. C., Panda, B., Tran, J. P. and Broeckhoven, C. (2021). Biomimicry for 3D concrete printing: A review and perspective. Additive Manufacturing. 38(2016), p. 101823. doi: 10.1016/j.addma.2020.101823

Dziadosz, A. and Kończak, A. (2016). Review of selected methods of supporting decision-making process in the construction industry. Archives of Civil Engineering. 62(1), pp. 111–126 doi: 10.1515/ace-2015-0055

Elmousalami, H. H. (2020). Artificial intelligence and parametric construction cost estimate modeling: State-of-the-art review. Journal of Construction Engineering and Management. 146(1), p. 03119008. doi: 10.1061/(asce)co.1943-7862.0001678

Fang, Y. and Ng, S. T. (2011). Applying activity-based costing approach for construction logistics cost analysis. Construction Innovation, 11(3), pp. 259–281. doi: 10.1108/14714171111149007

Hong, J., Shen, Z., Li, Z., Zhang, B. and Zhang, W. (2018). Barriers to promoting prefabricated construction in China: A cost–benefit analysis. Journal of Cleaner Production. 172(2018), pp. 649–660. doi: 10.1016/j.jclepro.2017.10.171

Igwe, U. S., Mohamed, S. F., Dzahir, M.A., Yusof, Z.M. and Khiyon, N.A. (2020). Towards a framework of automated resource model for post contract cost control of construction projects. International Journal of Construction Management, pp. 1–10. doi: 10.1080/15623599.2020.1841550.

Illankoon, I. M. C. S. and Lu, W. (2019). Optimising choices of "building services" for green building: Interdependence and life cycle costing. Building and Environment, 161, p. 106247.

Kim, K., Lee, G. and Kim, S. (2020). A study on the application of blockchain technology in the construction industry. KSCE Journal of Civil Engineering. 24(9), pp. 2561–2571.doi: 10.1007/s12205-020-0188-x

Kim, C., Park, T., Lim, H. and Kim, H. (2013). On-site construction management using mobile computing technology. Automation in Construction. 35(1), pp. 415–423. doi: 10.1016/j.autcon.2013.05.027

Kimoto, K., Endo, K., Iwashita, S. and Fujiwara, M. (2005). The application of PDA as mobile computing system on construction management. Automation in Construction, 14(4), pp. 500–511.

Koseoglu, O. and Nurtan-Gunes, E. T. (2018). Mobile BIM implementation and lean interaction on construction site: A case study of a complex airport project. Engineering, Construction and Architectural Management, 25(10), pp. 1298–1321. doi: 10.1108/ECAM-08-2017-0188

Magill, L. J., Jafarifar, N., Watson, A. and Omotayo, T. (2020). 4D BIM integrated construction supply chain logistics to optimise on-site production. International Journal of Construction Management, pp. 1–10. doi: 10.1080/15623599.2020.1786623

Omotayo, T., Awuzie, B. and Olanipekun, A. O. (2020). An artificial neural network approach to predicting most applicable post-contract cost-controlling techniques in construction projects. Applied Sciences, 10(15), pp. 5171. doi: 10.3390/app10155171

Orji, I. and Wei, S. (2016). A detailed calculation model for costing of green manufacturing. Industrial Management & Data Systems. Emerald Group.

Pan, W. and Sidwell, R. (2011). Demystifying the cost barriers to offsite construction in the UK. Construction Management and Economics. 29(11), pp. 1081–1099. doi:10.1080/01446193.2011.637938

Perera, A. J. and Imriyas, K. (2004). An integrated construction project cost information system using MS Access™ and MS Project™. Construction Management and Economics. 22(2), pp. 203–211. doi: 10.1080/0144619042000201402

Pilkington, M. (2016). Blockchain technology: Principles and applications. In F. X. Olleros & M. Zhegu (Eds.). Research Handbook on Digital Transformations. Edward Elgar Publishing.

Qian, X. and Papadonikolaki, E. (2020). Shifting trust in construction supply chains through blockchain technology. Engineering, Construction and Architectural Management. 28(2), pp. 584–602. doi: 10.1108/ECAM-12-2019-0676

Robichaud, L. B. and Anantatmula, V. S. (2011). Greening project management practices for sustainable construction. Journal of Management in Engineering. 27(1), pp. 48–57. doi: 10.1061/(asce)me.1943-5479.0000030

Sakin, M. and Kiroglu, Y. C. (2017). 3D Printing of buildings: Construction of the sustainable houses of the future by BIM. Energy Procedia. 134, pp. 702–711. doi:10.1016/j.egypro.2017.09.562

Turk, Ž. and Klinc, R. (2017). Potentials of blockchain technology for construction management. Procedia Engineering. 196, pp. 638–645. doi: 10.1016/j.proeng.2017.08.052

Wang, J., Wu, P., Wang, X. and Shou, W. (2017). The outlook of blockchain technology for construction engineering management. Frontiers of Engineering Management. 4(1), pp. 67–75. doi: 10.15302/j-fem-2017006

Wang, Y. R., Yu, C. Y. and Chan, H. H. (2012). Predicting construction cost and schedule success using artificial neural networks ensemble and support vector machines classification models. International Journal of Project Management. 30(4), pp. 470–478. doi: 10.1016/j.ijproman.2011.09.002

Wu, P., Wang, J. and Wang, X. (2016). A critical review of the use of 3-D printing in the construction industry. Automation in Construction. 68, pp. 21–31. doi: 10.1016/j.autcon.2016.04.005

Yaga, D., Mell, P., Roby, N. and Scarfone, K. (2019). Blockchain technology overview. *National Institute of Standards and Technology Internal Report.* arXiv:1906.11078

Zaher, M., Greenwood, D. and Marzouk, M. (2018). Mobile augmented reality applications for construction projects. Construction Innovation. 18(2), pp. 152–166. doi: 10.1108/CI-02-2017-0013

SECTION B
STRATEGIES

6 Overhead Cost Reduction and Maintenance through Continuous Improvement

Temitope Omotayo, Udayangani Kulatunga, and Bankole Awuzie

6.1 Introduction: The hidden costs of construction and continuous improvement

Scholars and practitioners have termed construction as an important economic activity, which makes use of a large percentage of a country's resources in the delivery and maintenance of buildings and associated infrastructure (Mu'azu, 2002). As a result, the construction industry is perceived to occupy an essential position for the development and growth of developing and developed economies. This is because the construction industry provides, maintains, and sustains infrastructure, which is essential for enhancing the quality of life of the citizenry (Ogunlana, 2010). In the same manner, Usman, Inuwa, Iro, and Dantong (2012) added that the construction industry facilitates the actualisation of socio-economic objectives such as infrastructure, shelter, and employment opportunities. Participants within this industry include consultants such as quantity surveyors, engineers, architects, clients, contractors, and subcontractors who are usually entities with differing goals, loyalties, and expectations (Dada & Akpadiaha, 2012; Oladirin, Olatunji, and Hamza, 2013). This is consistent with the argument put forward by Ajagbe, Ismail, Aslan, and Choi (2012) and Ismail, Tengku-Azhar, Yong, Aslan, Omar, Majid, and Ajagbe (2012), namely that organisations serve as a source of economic growth in the form of innovation, wealth, and job creation, irrespective of size. Hence, the construction industry brings different participants together to promote the economy of any country (Memon, 2014).

To promote economic and social development, construction projects must be successfully implemented to accomplish the desired levels of technical performance, maintain their schedule, and remain within budget (Ganiyu & Zubairu, 2010). Such relevance and the essence of developing effective strategies for managing costs within the construction industry is highlighted by Šiškina, Juodis, and Apanavičiene (2009, p. 215), who assert that "the only way to increase the company's competitiveness under highly intense competition in construction market with declining building contractors' profits and shrinking market shares is to control the costs of production and

DOI: 10.1201/9781003176077-8

business". This suggests that the effective and appropriate management of a company's overhead cost could constitute a starting point for successes such as profitability and client satisfaction. Hence, a too low or too high overhead cost due to inappropriate or poor evaluation of overhead costs will have a significant influence on the annual profit of construction companies.

Construction cost can be divided into direct cost, risks and profit, and overhead cost, which is also known as indirect cost (Shakantu, Tookey & Bowen, 2003; Šiškina et al., 2009). Construction overhead costs are referred to as hidden costs that cannot be quantified, unlike other construction cost elements extracted through the process of detailed measurement of drawings and rate build-up. Overhead costs are usually hidden in construction projects and may result from office expenses, building rentals, transportation costs, staff wages, social security, taxes, and insurance, just to mention a few (Šiškina et al., 2009). If the hidden overhead costs of construction are not clearly itemised, planned for, and controlled during the construction stage, the chances of cost escalation will be high. Consequently, there is a need to understand the nature and factors influencing construction overhead cost. According to Assaf, Bubshait, Atiyah, and Al-Shahri (2001), company and project overheads are two major types of overheads in construction. These types of categories of overheads are discussed in this chapter.

6.2 Continuous improvement of construction company's overhead cost

A construction company's overhead cost refers to all costs or expenses incurred by the company in developing support for production processes required to maintain its business (Adrian, 1982). These costs or expenses, or both, are not directly related to a specific job or project (Assaf et al., 2001). Also, these costs are referred to as general overhead costs (Peurifoy & Oberlander, 2002). This description is consistent with the views espoused by Dagostino and Feigenbaum (2003), who assert that company overhead costs are cost items that represent the cost of doing business. These expenses are perceived to be fixed in nature and are paid for, irrespective of whether the organisation is carrying out a job or not. These overhead costs are referred to as general costs because they include all those expenses incurred by the company which cannot be directly tied to a definite job or project.

Šiškina et al. (2009) concluded that a construction company's overhead cost only reflects its management system, use of its available assets, and organisation of its activity. This view suggests the need for construction companies to know their true overhead costs to facilitate appropriate cost recovery. Arguably, the inability of construction companies to manage their overhead costs effectively is a major contributor to their lack of sustenance and poor financial performance.

Šiškina et al. (2009) identified company overheads as charges associated with the management of personnel, administrative charges related to payments, expenses for the maintenance of buildings and premises, and the size of building facilities. Other overheads include the rental of company buildings and assets and the cost of utilities or clerical duties. Such expenses are commonly distributed among the company's jobs or projects. Company overhead costs are associated with indirect costs incurred because of the daily operations of an organisation. Assaf et al. (2001) asserted that company overhead costs are difficult to quantify for the following reasons:

- They are not clearly defined.
- They are essential for the operations of the company and any reduction may influence quality.
- They are not accurately identified, and financial allocations are necessary.
- Reduction in company overhead costs can affect top management and excessive overhead cost may be blamed on other causations.
- Unclear plans for company overhead costs may lead to higher costs.

Construction company overhead costs usually range between 8% and 15% (Anass, Elfezazi, Hurley, Garza-Reyes, Kumar, Anosike, and Batista, 2017; Assaf et al., 2001; Liu, Wu, Yue, & Zhang, 2019). The factors affecting the allocation of a percentage for a company's overhead costs vary and may be influenced by variables such as the geographical location of the company, specialisation, the nature of top management, the experience of the contractor, company policy on overheads, the type of contractor, the payment schedule, the financial position of the company, the size of the construction company, competition, and the scale of construction projects. Construction companies' overhead costs are not fixed and can always be adjusted in response to the financial requirement and employers' expectations.

The management of overhead costs remains vital for efficient construction company operations. However, in many construction companies, overhead costs may create unexpected financial losses if they are not managed effectively. As such, the continual reduction of all overhead costs to an optimised level is vital. According to Omotayo, Olanipekun, Obi, and Boateng (2020), non-physical wastes are generated from activities such as time wastage or unproductive activities, which may lead to higher overhead costs. Overhead costs cannot be avoided but can be reduced to a level that guarantees profit attainment for the company.

The importance of incorporating continuous improvement practices in construction companies' overhead costs management can be found in the concept of gemba kaizen. Gemba kaizen is continuous improvement in the workplace. Anass et al. (2017) and Imai (1992) confirm that gemba kaizen can be used as a continuous improvement approach in the place of work to standardise operations. Organisational business ethics, organisational waste reduction policies, and the incorporation of continuous improvement policies are essential to achieving gemba kaizen in construction companies (Omotayo & Kulatunga, 2017). Similarly, continuous improvement consciousness in construction companies can pave the way for improved overhead cost management.

The management and continuous improvement of company overhead costs also depend on several factors. These factors were identified by Assaf et al. (2001) as including the following:

- Delayed payments
- Absence of new projects
- Inflation rate
- Government regulations
- Construction company's growth
- Employer's related reasons
- Fluctuation and market forces

Apart from a construction company's growth, the abovementioned factors are mostly exogenous, hence making it difficult to implement continuous improvement since they consist of circumstances that are beyond the control of top management. Therefore, overhead costs may increase when the construction company's operations are impacted by unfavourable external forces. Equally, construction project overhead costs can determine the percentage allocation of a company's overhead costs.

6.3 Continuous improvement of construction project overhead cost

Construction project overheads are cost items, which are not labour, materials, or production equipment that can be identified with a project or job. These costs are incurred in the course of carrying out a job or implementing a project. Dagostino and Feigenbaum (2003) define these costs as consisting of expenses that are required to construct a project or job as opposed to being charged directly to a branch of work. Bauer, Koppelhuber, Wall, and Heck (2017) identified costs such as construction manager's cost, specific salaries of unproductive project-related staff such as cleaning and security staff, and interest and depreciation of equipment on the construction site as project overhead costs.

Indirect costs which cannot be quantified or determined easily at the point of producing construction cost information can be categorised under overhead costs. Construction project overhead costs can be derived from preliminary items of work such as the following:

- Insurance cost
- Cost of electricity
- Water supply

- Scaffolding
- Transportation
- Access roads
- Temporary fencing
- Temporary office site
- Cost of security
- Site administrative charges
- Telephone or mobile communication cost
- Construction staff welfare
- Furniture
- Consumables
- Protective clothing
- Plant and equipment maintenance cost
- Setting out
- Site waste clearance

However, preliminary costs in construction are usually distinguished from overhead costs when including a percentage. The term "markup" covers preliminary costs and overhead and profit costs, which constitute an integral part of the final construction cost. In a different markup approach, preliminary costs ranging between 8% and 12% may be included separately as part of the construction cost. Construction project overhead costs also depend on the nature of the project. For instance, a bridge construction project will have its own unique preliminaries when compared with a building's construction cost. Therefore, it is evident that the nature of construction overhead costs is dependent on several factors.

Studies conducted by Chan (2012), Leśniak and Juszczyk (2018), Liu et al. (2019), and Shakantu et al. (2003) itemised influential drivers for construction cost overheads. Table 6.1 provides a list of drivers influencing the incidence of higher or lower construction project cost overheads.

Following from Table 6.1, a classification of these drivers into six categories can be discerned. These categories include site conditions, construction design economics, construction planning and procurement arrangements, additional costs, economic conditions, and design. These categories are briefly discussed below.

6.3.1 Site conditions

The location of a construction site will impact the nature of overheads that may be incurred on that project. For instance, a construction project situated on a swampy site will have groundwater and sand-filling-related costs. Likewise, a bridge construction over water-logged sites will incur cost associated with the construction of cofferdams. Similarly, construction projects in a conflict zone will incur more overhead costs when compared to those in peaceful regions. Other activities related to the nature

Table 6.1 Factors influencing construction project overheads

S/N	Factors	Drivers
1	Site conditions	Location of construction site
		Nature of construction logistics
		Shape of site
		Soil conditions
2	Construction design economics	Gross floor area of the project
		Project construction estimate
		Nature of project
		Schedule of project
		Proposed building height
		Shape of structure
3	Construction planning and procurement arrangements	Construction information management
		Construction methods
		Quality requirements
		Contractor selection method
		Procurement strategy
		Nature of contract
		Payment arrangement
		Input of contractor on design
		Bond or warranty requirements
		Extent of adopting BIM
		Community engagement
		Payment delays
		Managing subcontractor
		Variations and rework
4	Additional costs	Plant and equipment maintenance
		Fuel price
		Environmental cost
		Plant depreciation cost
		Material handling and delivery
		Inflation
5	Economic condition	Economic forces
		Economic of site location
		Construction industry output
		Interest rates
		Exchange rates
6	Design	Constructability
		Incomplete design
		Errors in design

of the site which may influence overhead costs include the nature of construction logistics, such as setting out equipment, and plant requirements. Also, some form of bulk excavations, removal of rock boulders and obstruction may be required for some construction sites. The topography or shape of the site and attendant soil conditions will influence the incidence of overhead costs. If the terrain of a site is undulating, the decision to excavate or fill specific sections of the site will add to any existing overhead costs.

6.3.2 Construction design economics

Design economics pertains to the concept of designing to fit the cost of employers. Every construction project has a cost limit before the development of the budget. In deriving a suitable cost for designed spaces, the less the gross floor area of a construction project, the less the overhead costs will be. A smaller gross floor area will provide fewer overhead costs. Conversely, the scale of construction projects will also impact the incidence of overhead costs. Minor, major, and megaprojects will individually have their own variable overhead costs associated with size or magnitude. Therefore, the larger the construction project, the more overhead costs will be provisioned for in the contract sum. Additionally, the shape and height of buildings will attract varying overhead costs. For instance, it is easier to construct rectangular and square buildings than circular buildings. Simple buildings will have fewer overheads as compared with complex building projects.

6.3.3 Construction planning and procurement arrangements

The procurement arrangement in a project may impact the percentage of overheads included final contract sum. For instance, a design-and-build procurement method will not make use of a bill of quantities and will therefore have more cost information in the schedule and bill of approximate quantities. Also, the design-and-build method may give rise to higher overhead costs as opposed to the traditional procurement method, where there is a complete bill of quantities, cost, and increased levels of price certainty. Consequently, construction planning goes a long way in determining cost and price certainty. For instance, if a construction project has been planned to include collateral warranties for subcontractors, a lump sum payment arrangement and the nature of the contract may all affect the percentage allocation of the overhead costs. As part of construction planning, decisions to adopt new technology for stakeholder engagement and information management may lower or raise the overhead costs. In instances where BIM has been implemented for construction procurement and information management, overhead costs can be minimised by avoiding costs associated with non-value-adding activities such as paper printing and decentralised information distribution.

6.3.4 Additional costs

Additional construction costs arising from variations and rework can contribute to overhead costs. Where errors have been detected in drawings, mistakes made during construction, or a misinterpretation of construction contract documents occurs, additional costs for rework will increase construction project overhead costs. This type of construction project overhead cost is unplanned for but yet remains essential for project delivery. Hence, overhead costs are unavoidable and must be factored into the overall construction cost. Additional costs of maintaining plant, equipment, and depreciation costs are important in calculating construction costs. Another important construction project overhead cost results from material handling and delivery. In the rate build-up, it is important to always include wastage in the estimated material handling and delivery costs. Although overhead costs are different from the percentage inclusion for material handling and delivery costs, it has been observed that during construction activities where variations and rework activities are included, such costs associated with material handling and delivery will increase.

6.3.5 Economic conditions

Economic conditions generally may affect overhead costs in construction planning and projects. Price fluctuations resulting from inflation indices, economic factors of the country, and location of the site may also raise or lower overhead costs. In predicting overhead cost of construction, interest rates are expected to be stable. Nonetheless, in countries with volatile interest rates, construction costs may experience instability because of excessive overhead costs. Fluctuations in exchange rates in construction projects where most materials are imported from other countries may influence overhead costs negatively or positively and inevitably also construction costs.

6.3.6 Design

Constructability pertains to the ease of transference or interpretation of design into real-life construction. There have been instances where construction designs are difficult to build, not because of the cost but owing to the nature of the design. For instance, it may be difficult to construct a circular building because of the shape and technicalities involved in the process. Rectangular or square-shaped buildings are easier to construct when compared with circular buildings. Furthermore, constructability may have a major impact on the incidence of overhead costs. If the design is incomplete or too complex, overhead costs may be higher. Errors in construction designs can lead to variations and rework. Variations and rework, as discussed in Section 6.3.4, affect the level of overhead costs.

6.4 Basis of calculating overhead costs in construction

In determining overhead costs in the construction industry, constituents of indirect costs, excluding materials costs, are used as a basis for calculating overhead costs. Scevik and Vitkova (2017) postulated a formula for calculating overhead base and the volume of overhead cost (R) as follows:

$$R = (Rv + Rs) \qquad\qquad (6.1)$$

Rv is given as production overhead costs
Rs is the administrative overhead costs

Consequently, overhead surcharge (Rp) is calculated according to the formula:

$$Rp = R/Rz * 100 \qquad\qquad (6.2)$$

Rp is an overhead surcharge (in per cent) R represents the volume of
    overhead costs
Rz is the volume of the overhead costs

Although some contractors may have an in-house approach to calculating overhead costs, the cost determinants are based on the nature of the construction project, the financial projections for profit, the subcontractors' cost, payment arrangements, inflation, the exchange rate, the availability of building materials and other factors itemised in Table 6.1.

Overhead cost calculations may be determined by taking a percentage of the contract sum and adding it back to the contract sum. Contractors may also provide detailed estimating of construction preliminary items of works which generally contribute to overhead costs by calculating the breakdown of cost estimated for each item.

6.5 Conclusion: Strategies for overhead cost maintenance and reduction

Construction projects often experience cost distortion (Cokins, 1996; Horngren et al., 1999; Johnson & Kaplan, 1987). This is because overhead costs are assigned to work divisions in proportion to direct labour costs or direct labour hours (Sommer, 2001). These views suggest that companies do not know the actual overhead cost for each participant such as subcontractors or work division. Overhead costs stem from preliminary items of works and activities which largely contribute to the construction process. Conversely, these costs cannot be quantified easily. Shakantu et al. (2003) noted that increased utility of fixed assets such as vehicles, roads, and transportation cost in material delivery may lead to higher overheads and subsequently higher construction cost. Additionally, Omotayo et al. (2020) elucidated the strategies for continuous improvement in the construction industry and the continual elimination of waste through overhead costs reduction. Omotayo et al. (2020) identified the following aspects of construction overheads as essential for engendering a reduction in overhead costs:

- *Continual reduction of plant and equipment depreciation overhead cost throughout the construction phase will keep the project cost within budget*

  Overhead costs associated with plant and equipment depreciation cost may be minimised by adopting new construction technologies such as 3D printing and offsite construction practices. New construction technology will reduce or eliminate overhead costs which are tied to construction plants and equipment. Depreciation cost is a major contributor to overhead costs and by adopting sustainable construction technologies, construction costs can be managed effectively.
- *Continual cost reduction of overhead cost of activities related to mobilisation and equipment setup will keep the project cost within budget*

  Mobilisation and equipment setup in construction sites make up indirect costs that may not be factored into construction cost. Therefore, it is important to identify all preliminary items of works that are synonymous with construction site mobilisation and equipment set up before commencing construction sites. 5D BIM can be applied to identify and map all preliminary costs in the final level of design development from the contractor's perspective.
- *Continual reduction of activities related to drawing reviews and other variations or alterations will eliminate unnecessary cost, thereby keeping the project cost within budget*

  Overhead costs come from drawing reviews, variations, and rework. Error correction can benefit from modern methods of construction and new approaches to design and construction such as real-time collaboration, supply chain integration with the procurement systems, and the application of a construction project management software.
- *Ensuring activities related to construction variations are continually minimized will create more profit for the contractor*

  One of the easiest ways of ensuring continuous improvement in construction cost is by reducing non-value-adding activities for profit maximisation. From the perspective of the contractor, profit is important, likewise value and client satisfaction. Overhead costs are indirect costs, some of which are activities that do not add value directly to construction processes and must be identified and mitigated.
- *Cost of activities related to purchase orders and material deliveries can be reduced continually throughout the construction phase to control the project cost for optimum profit*

  In attaining expected profits in construction, purchase orders and material deliveries have hidden costs such as administrative charges, multiple orders, and transportation costs. If there is a conscious identification of hidden costs linked with the purchase and delivery of materials, overhead costs will be reduced and the chances of reaching the profit benchmark may be realised.

- *Overhead cost related to paying suppliers, subcontractors and labourers can be reduced continually throughout the construction phase to keep the project cost within budget*

  Payment of suppliers, subcontractors, and onsite labourers have their own hidden costs, which add up to form overheads. The application of a construction payment software, cloud-based computing, and blockchain technology may reduce hidden costs of payments to suppliers, subcontractors, and labourers. Blockchain technology and cryptocurrency adoption in the construction industry are still very new. However, the application of cryptocurrency digital wallets in construction payments can avoid any hidden fees associated with payments.

- *Continual reduction of overhead costs related to construction cost planning, general planning, resource planning, and project reports will create more profit for the contractor*

  Construction planning and report production incur overhead costs which remain unknown to the construction planner and project stakeholders. Every construction project is unique, and it is the prerogative of construction project planners and estimators to identify all forms of hidden costs linked with construction planning and report writing activities. This may be achieved by adopting a clear framework for documentation and information management peculiar to the project needs.

- *Continual reduction of overhead costs associated with preliminary items of work such as site office, storage, security, electricity, water supply, first aid, and the like will eventually help the creation of more profit and improve project delivery*

  Preliminary items of work generally contribute to overhead costs because they are mostly unquantifiable and there is no specific pattern for estimating individual works on site. Likewise, project complexity, procurement approaches, and contractual arrangement can make preliminary costs almost impossible to estimate. In this regard, to reduce overhead cost continually, it is strategically vital to rely on historical overhead cost information of preliminary items of work for similar construction projects and designs. The estimation of construction preliminary costs using the rebasing methods may provide some clarity on the nature of preliminary items if overhead costs are to be understood in a project.

Overhead costs cannot be eliminated in construction projects. Conversely, overhead costs should not be too high and must be investigated as part of construction project cost estimation. The use of "overheads and profit percentage" inclusion in construction costs has not been helpful in mitigating construction cost escalations. Consequently, the process of identifying overhead costs in construction projects may benefit from the types of previous overheads in completed construction projects. Overhead costs can be eliminated from construction projects through the adoption of new technologies such as offsite construction where onsite labour and plants will be reduced, BIM where errors will be minimised in designs, and blockchain payment methods for the elimination of hidden payment fees. Continuous improvement in construction cost can be attained through overhead cost maintenance or reduction process and the best strategy for realising objective is tied to the complete espousal of modern methods of construction.

References

Adrian, J. J. (1982). *Construction estimating.* Reston, VA: Reston Publishing Company.

Ajagbe, A. M., Ismail, K., Aslan, S. A. and Choi, L. S. (2012). Investment in technology-based small and medium-sized firms in Malaysia: Roles for commercial banks. *International Journal of Research in Management and Technology*, 2(2), pp. 147–153.

Anass, C., Elfezazi, S., Hurley, B., Garza-Reyes, J. A., Kumar, V., Anosike, A. and Batista, L. (2017). Green and lean: A Gemba–Kaizen model for sustainability enhancement. *Production Planning and Control*, 30(5–6), pp. 385–399. doi: 10.1080/09537287.2018.1501808

Assaf, S. A., Bubshait, A. A., Atiyah, S. and Al-Shahri, M. (2001). The management of construction company overhead costs. *International Journal of Project Management*, 19, pp. 295–303.

Bauer, B., Koppelhuber, J., Wall, J. and Heck, D. (2017). Impact factors on the cost calculation for building services within the built environment. *Procedia Engineering*, 171, pp. 294–301.

Chan, C. T. W. (2012). The principal factors affecting construction project overhead expenses: An exploratory factor analysis approach. *Construction Management and Economics*, 30(10), pp. 903–914. doi: 10.1080/01446193.2012.717706.

Chan, E. H. W. and Lee, G. K. L. (2008). On a sustainability evaluation of government-led urban renewal projects. *Journal of Facilities*, 26(13), pp. 526–541.

Cokins, G. (1996). *Activity-based cost management making it work: A manager's guide to implementing and sustaining an effective ABC system.* Burr Ridge, IL: Irwin Professional Publishing.

Dada, M. O. and Akpadiaha, B. U. (2012). An assessment of formal learning processes in construction industry organisations in Nigeria. *International Journal of Architecture, Engineering and Construction*, 1(2), 103–111.

Dagostino, F. R. and Feigenbaum, L. (2003). *Estimating in building construction* (6th ed.). Upper Saddle River, NJ: Pearson Education.

Ganiyu, B. O. and Zubairu, I. K. (2010). Project cost prediction model using principal component regression for public building projects in Nigeria. *Journal of Building Performance*, 1(1), p. 2010.

Horngren, C. Y., Foster, G. and Datar, S. M. (1999). *Cost accounting* (10th ed.). Upper Saddle River, NJ: Prentice Hall.

Imai, M. (1992). Comment: Solving quality problems using common sense. *International Journal of Quality & Reliability Management*, 9(5), pp. 71–75. doi: 10.1108/EUM0000000001655

Ismail, K., Tengku-Azhar, N. T., Yong, Y. C., Aslan A. S., Omar, Z. W., Majid, I. and Ajagbe, M. A. (2012). Problems on commercialization of genetically modified crops in Malaysia. *Procedia – Social and Behavioral Sciences*, 40, pp. 353–357.

Johnson, T. and Kaplan, R. (1987). Rise and fall of management accounting. *Management Accounting, IMA*, 1, pp. 22–30. https://www.hbs.edu/faculty/Pages/item.aspx?num=2609

Leśniak, A. and Juszczyk, M. (2018). Prediction of site overhead costs with the use of artificial neural network based model. *Archives of Civil and Mechanical Engineering*, 18(3), pp. 973–982. doi: 10.1016/j.acme.2018.01.014.

Liu, S., Wu, J., Yue, Y. and Zhang, Y. (2019). Analysis of factors affecting the hidden costs of construction projects factor analysis. *Open Access Library Journal*, 6(8), 1–8.

Memon, A. H. (2014). Contractor perspective on time overrun factors in Malaysian construction projects. *International Journal of Science, Environment and Technology*, 3(3), pp. 1184–1192.

Mu'azu, D. A. (2002). The role of the professional builder in the Nigerian constructions industry. *ATBU Journal of Environmental Technology*, 1(1), pp. 29–31.

Ogunlana, S. O. (2010). *Sustaining 20:20:20 vision through construction: A stakeholder participatory approach.* Second Distinguished Lecture in the Distinguished Lecture Series of the School of Postgraduate Studies, University of Lagos, Lagos, Nigeria.

Oladirin, O. T., Olatunji, S. O. and Hamza, B. T. (2013). Effect of selected procurement systems on building project performance in Nigeria. *International Journal of Sustainable Construction Engineering and Technology*, 4(1), pp. 48–62.

Omotayo, T. and Kulatunga, U. (2017). A gemba kaizen model based on BPMN for small- and medium-scale construction businesses in Nigeria. *Journal of Construction Project Management and Innovation*, 7(1), pp. 1760–1778.

Omotayo, T., Olanipekun, A., Obi, L., Boateng, P. (2020). A systems thinking approach for incremental reduction of non-physical waste. *Built Environment Project and Asset Management*, 10(4), doi: 10.1108/BEPAM-10-2019-0100.

Peurifoy, R. L. and Oberlander, G. D. (2002). *Estimating construction costs.* New York: McGraw-Hill.

Scevik, V. and Vitkova, E. (2017). Optimization of overhead costs of a construction contract. In Potocan, V.,Kalinic, P. and Velutic, A. (Eds.). In *Proceedings of the 26th International Scientific Conference on Economic and Social Development*, Zagreb, pp. 515–523.

Shakantu, W., Tookey, J. E. and Bowen, P. A. (2003). The hidden cost of transportation of construction materials: An overview. *Journal of Engineering, Design and Technology*, 1(1), pp. 103–118. doi: 10.1108/eb060892.

Šiškina, A., Juodis, A. and Apanavičiene, R. (2009). Evaluation of the competitiveness of construction company overhead costs. *Journal of Civil Engineering and Management*, 15(2), pp. 215–224. doi: 10.3846/1392-3730.2009.15.215-224.

Sommer, B. (2001). *Personal communication.* Redwood City, CA: DPR Construction, Inc.

Usman, N. D., Inuwa, I. I., Iro, A. I. and Dantong, J. S. (2012). Training of contractors' craftsmen for productivity improvement in the Nigerian construction industry. *Journal of Engineering and Applied Science*, 4, pp. 1–12.

7 Global Construction Organisational Cultures and Continuous Improvement

Temitope Omotayo, Udayangani Kulatunga, and Bankole Awuzie

7.1 Introduction: Construction organisational culture

In the process of implementing continuous improvement in the construction industry, construction organisations must modify their culture to fit the concept of continuous improvement. Consequently, a national organisational culture of construction companies is peculiar to the nature of construction projects they engage in, the geographical location, the government policies, the organisational policies, and mission statements. Omotayo, Awuzie, Egbelakin, Obi, and Ogunnusi, (2020) opined that for continuous improvement to be implemented in the construction industry, construction organisations must align their organisational culture with the ideals of waste reduction or minimisation, enshrine the elimination of non-value adding activities in the workplace through the concept of *gemba kaizen*, and adopt the Deming's circle in their management strategy. The nature of construction organisational culture is reviewed from various perspectives in this chapter for a clearer understanding of how continuous improvement can be established as a core notion of construction organisational growth.

7.2 Influence of organisational culture on change in the construction industry

There are many complexities associated with organisational culture. In the construction industry, these complexities pervade the dimensions of endogenous factors such as employer–employee relationship management, communication, finances, organisation growth, vision, and mission statement. Exogenous factors such as the construction professional regulatory bodies, government influences, economic climate, inflation, scale of construction projects, and demands of the construction economy all contribute to the formation of an organisational culture.

Hofstede (1984, p. 21) described organisational culture as "a collective programming of the mind which distinguishes the members of one organisation from another". Hence, within an organisation, there are distinct

DOI: 10.1201/9781003176077-9

individual cultures that are expected to form a singularity of purpose. Hofstede (1984) also recognised individualism in collectivism as dimensions for appraising the difference in national cultures between IBM managers.

Dawson (1997) described organisational culture as a convergence of distinct values and beliefs with individual thought processes within an organisation. Dawson's perspective of organisational culture aligns with Hofstede's definition and they both agree that employees in an organisation contribute to the ideals of an organisation.

From another viewpoint, organisational culture is a combination of attributes that may be delineated by the purpose (Jaeger, Yu, & Adair, 2017; Newman, 2007). Construction organisational culture can be studied from the perspective of culture type and dimension. Jaeger et al. (2017) noted that culture dimensions are organisational power, role, tasks, and persons. The dimension of organisational culture depends on the socio-psychological, technological, and social structural perspective (Liu, Shuibo, & Meiyung, 2006). The study of construction organisational culture is committed to achieving effective change towards continuous improvement. This is because some organisations already have a culture of openness to innovations and new concepts such as continuous improvement. Considering the varying approaches to studying organisational culture in the construction industry, national perspectives of organisational culture globally influence the way in which change is perceived and accepted. Cameron and Quinn (1999) identified four (4) main categories of organisational culture. They are hierarchical, market, clan, and adhocracy. Maiti and Indhu (2018) further suggested that these categories may be external and internal focus, stability and control, and flexibility and discretion organisational cultures. This chapter makes use of the organisational culture assessment instrument (OCAI) developed by Cameron and Quinn (1999) as a tool for understanding the perception of continuous improvement in organisations. The OCAI is a widely used and studied tool in management studies and it provides a streamlined approach to studying organisational cultures from national standpoints.

In providing a clearer understanding of how organisational culture influenced the direction and acceptance of continuous improvement in the construction industry, the subsequent sections address organisational culture from all six (6) continents, including Asia, Africa, Australia, Europe, and North and South America.

7.3 Asian construction organisational cultures

The largest and unique Asian construction economies such as those of China, India, and Saudi Arabia have differing construction organisational cultures.

7.3.1 China

Zhang and Liu (2006) studied the organisational culture of Chinese contractors in terms of culture-effectiveness modelling. Findings from that study showed that hierarchy and clan cultures dominate Chinese construction companies. These findings were corroborated by the findings from a similar study by Liu et al. (2006). According to Liu et al. (2006), the hierarchical culture remained common in the Chinese construction industry. A hierarchical culture in an organisation exhibits a formalised workplace with guidelines and procedures to ensure long-term stability and execution of tasks. In clan cultures, individual activities have commonalities that are shared among themselves. The clan ideals in an organisation provide an inclusive environment fostering teamwork and continuous improvement. The hierarchical culture of Chinese construction companies may limit innovation; however, it has proved essential in engendering their growth and development. On the other hand, the clan culture promotes openness towards continuous improvement.

7.3.2 India

In terms of the Indian construction organisational culture, Maiti and Indhu (2018) studied the organisational culture of several Indian construction organisations practising within that geographical context. Their findings revealed a prevalence of the market culture more than a clan or hierarchical culture. This genre of organisational culture is based on completing tasks and achieving outcomes. A market culture exhibits the attributes of targets, individualism instead of collectivism, and high expectations to succeed motivates organisations. This approach towards organisational culture may seem to negate innovation. This attitude may be marked with an unstructured approach to exploring opportunities presented through innovation. On the other hand, there is a perceived positive approach towards adopting any winning strategy that will provide a competitive edge.

7.3.3 Saudi Arabia

Saudi Arabian construction organisations are based on clan culture, driven by religious precepts amid challenges attributed to nepotism (Aldraehim, Edwards, Watson, & Chan, 2012; Al-Sedairy, 2001). Religion plays a significant part in the organisational culture adopted by construction organisations in Middle Eastern countries. Aldraehim et al. (2012) expounded on the challenges of Saudi Arabian organisations where clan culture is regarded as being tribal, Islamic, and kinship-related in nature. The attributes negatively influenced the workplace and desire to innovate. It can be inferred that there is a special clan culture of Saudi Arabian construction organisations, placing importance on family norms and values as directed by Islamic religious laws. In so doing, the opportunities for accepting new concepts and innovation are very limited.

7.4 African construction organisational cultures

Nigeria, South Africa, and Egypt represent the largest construction economies in Africa. This justified their selection as contexts within which to study the prevalent organisational culture dimensions of construction organisations operating there and the impact of such dimensions on their innovative potential vis-à-vis continuous cost improvement adoption and implementation. This section briefly highlights the nature of organisational cultures in these countries.

7.4.1 Nigeria

In the Nigerian construction industry, organisations have been shown to display a market culture (Alao & Aina, 2020). Other attributes of Nigerian construction organisations where the market culture exists are communication, employee engagement, and adaptability (Alao & Aina, 2020), thereby merging the concepts of the market culture, which is competition and goal orientation, with the dynamics of improved employee–employer relationship management. Abiola-Falemu, Ogunsemi, and Oyediran (2010) observed that smaller construction companies in Nigeria have more of a culture to innovate than larger ones. This desire to innovate in smaller construction companies may be attributed to the desire to compete with larger construction organisations. The construction companies also exhibit a power culture that focusses on external maintenance with the desire for control and stability. A power culture is a hybridised version of a hierarchy-market culture (Abiola-Falemu, Ogunsemi, & Oyediran, 2010). The power culture views external conditions as aggressive conditions which may displace their stability and influence their profit margins. The power culture is more synonymous with the hierarchy culture than internal conditions of the organisation.

7.4.2 South Africa

Similar to what pertains to Nigerian construction organisations, the market culture appears dominant in South Africa (Dlamini, 2015; Harinarain & Bornman, 2013). In addition to the market culture of construction organisations in South Africa, relationships among key stakeholders are effectively managed to enhance competitiveness and collaboration (Harinarain & Bornman, 2013). Also, the working environments are harmonious to reduce conflicts. Accordingly, the drive to adopt and implement innovative practices in the South African construction industry depends on the financial and intrinsic benefits such practices may offer. The predominant market culture is infused with relationship management and an increased desire to become more competitive.

7.4.3 Egypt

The organisational culture of Egyptian construction firms bears a close semblance to firms in Nigeria and South Africa (Dajani & Mohamad, 2017; El-Nahas, Abd-El-Salam, & Shawky, 2013). Additionally, the market organisational culture in Egypt is predicated on organisational transactional and transformational leadership approaches (Dajani & Mohamad, 2017). The market culture in Egyptian construction organisation is driven by leadership with a vision hinged on profitability. Hence, in some instances, this version of market culture may have some hierarchical attributes, thereby resembling the power culture. African construction organisations largely hinge on the market culture to grow financially and expand from their competitive advantage. However, there are many construction economies in Africa that depend on the nature of government policies; therefore, the market culture of these construction organisations is influenced by political and economic vagaries.

7.5 Australian construction organisational cultures

In this section, the organisational cultures of Australia and New Zealand are reviewed.

7.5.1 Australia

Gray, Densten, and Sarros (2003) described the Australian construction organisational culture as being supportive and innovative in nature. These attributes are aligned to the adhocracy culture whereby employees are allowed to share their ideas with the organisation and take risks for the purpose of innovation. This is a dynamic and innovative workplace. The pioneers or the officials are viewed as innovators and risk-takers. The long-term goal is to grow and develop new assets for the organisation. The organisation promotes individual initiative and freedom. Cheung, Rowlinson, Spathonis, Sargent, Jones, Jefferies, and Foliente (2004) reported that there may be information flow and communication issues in Australian organisations; however, relationship management remains a major attribute. Cheung et al. (2004) further noted that the adhocracy culture is deemed to be more important in Australian organisations. The role culture depends more on relevance to professionalism which is more important than innovation. Hence, the role culture is synonymous with bureaucratic and procedural systems. The role culture is very similar to the hierarchical culture, and it may be more prominent in larger organisations than smaller ones.

7.5.2 New Zealand

Clan and hierarchical organisational cultures are very dominant in the New Zealand construction industry (Frey, Boyd, Foster, Robinson, & Gott, 2016). Based on the results from a survey conducted by Frey et al. (2016), it was established that clan and hierarchical cultures were preferred by the managers working within the New Zealand context. However, the clan culture of fostering relationships within the organisation was discovered to be the leadership and staff management style of New Zealand organisations. Similarly, Durdyev and Mbachu (2011) discussed on-site productivity issues and the influence of organisational culture on productivity. Organisational culture was found to be very influential on project performance and delivery in New Zealand. Durdyev and Mbachu (2011) opined that New Zealand's cultural heritage values have a significant influence on the construction industry. This implies that there are several external cultural influences on existing organisational cultures, which may make New Zealand's organisational culture more clan-like in nature than hierarchical.

7.6 European construction organisational cultures

European construction economies are mainly dominated by the United Kingdom (UK), Germany, and France. As such, evidence drawn from the literature concerning the organisational cultures of construction firms situated in these countries is reviewed.

7.6.1 The United Kingdom

Von Meding, Kelly, Oyedele, and Spillane (2012) and Nazarian, Atkinson, and Foroudi (2017) studied the prevalent corporate and management culture of the UK construction industry. Results from those studies showed that the construction organisations in the UK construction industry were inclined more towards the clan, market, and adhocracy cultures. However, the studies also showed that the latter, which drives innovation, remains the least option while the market culture was most dominant. Hence, construction organisations in the UK prefer to react to external changes and demands for construction outputs. The UK construction companies are driven by profit and the desire to meet their financial targets. Von Meding, McAllister, Oyedele, and Kelly (2013) also noted that stakeholder management and corporate culture are crucial for the success of organisational culture and construction productivity. Relationship management is an important feature of UK construction organisations and it is a major approach to meeting the profit-driven construction industry.

7.6.2 Germany

The German construction organisational culture was labelled as adhocratic in nature (Engelen, Flatten, Thalmann, & Brettel, 2014). The adhocracy culture of German construction organisations may be linked with their national values of innovation, entrepreneurialism, originality, risk-taking, and openness to change (Felipe, Roldán, & Leal-Rodríguez, 2017). Construction organisations react to external influences rapidly by innovating and accepting change. The adhocracy culture in German construction organisations has the main characteristics of flexibility and management of information overload with creativity. There is freedom to express individual innovation while the acceptance of new concepts such as continuous improvement in the construction industry is feasible.

7.6.3 France

In France, construction organisational cultures exhibit homogeneity, which is closely related to the clan culture of organisations (Calori & Sarnin, 1991; Seibel, 1990). French organisational cultures showed less of the market and adhocracy cultures upon investigation. However, the clan culture in France cannot be generalised for all construction organisations. Calori and Sarnin (1991) also observed that there may be a potential for hierarchical culture in France, which implies a culture of order and bureaucracy. This indicates that the organisational culture of France is a mixture of clan and hierarchical features.

Other European countries such as Spain and Finland exhibit a mixture of adhocracy and hierarchical cultures (Felipe et al., 2017). Croatia was found to exhibit the clan culture, just like France (Strossmayer, Matotek, & Strossmayer, 2014). Across other European countries, there is a desire to innovate; therefore, creativity in organisations has become very common. As such, the adhocracy culture was becoming increasingly common.

7.7 North American construction organisational cultures

The North American continent is mainly dominated by the United States of America (USA) and Canada. This section considers the types of organisational in these countries.

7.7.1 The United States of America

In the USA, construction organisations express the clan culture whereby teamwork and employee development collectively provide a shared belief and vision (Felipe et al., 2017). Cameron and Quinn (1999) further described the clan culture as being necessary when organisations cannot respond to external influences or plan effectively for the future. The focus of the managers can then be directed to the employees of the organisation to collectively advance the goals of the organisation. Other organisational features such as strategic emphasis, leadership, criteria for success measurement, and general management are enshrined in the perspective of the clan culture.

7.7.2 Canada

O'Donnelle and Boyle (2008) suggested that the Canadian public administration and organisational culture is very similar to those of the UK and USA. With this similarity, Canadian organisations display features of the clan and market cultures. Canadian organisations also place more importance on value and behaviour development towards leadership and competence. The similar clan culture of the USA is common in Canada and there is a perceived notion of organisational growth through employee development. Dastmalchian, Lee, and Ng (2000) studied the interplay between organisational and competing national cultures in Canada. They found that there is consistent information regarding individualism in Canada. This individualism is a main feature of the clan culture where individual employees are the focus of organisations. The clan culture in organisations may be a positive model for continuous improvement in construction organisations. However, there must be a clear approach towards accepting individual innovations and creativity.

7.8 South American construction organisational cultures

The South American continent is largely dominated by Brazil, Chile, and Argentina. This section covers construction organisational cultures in these three countries.

7.8.1 Brazil

Thomas, Marosszeky, Karim, Davis, and McGeorge (2002) studied the organisational culture of Brazilian construction companies. The authors identified clan culture as a key feature. From an in-depth perspective, Ankrah and Proverbs (2005) described the organisational culture in Brazil as an intuitive, common sense, and experience-based system. This description is reminiscent of the clan culture where there is a focus on the internal constituents of their organisation rather than the external environment. Also, Ankrah and Proverbs (2005) noted that there are very few features of informed decision-making which are based on the financial measures and competitive environment.

7.8.2 Chile

Rodriguez and Stewart (2017) described the relationship between managers and workers in construction organisations as problematic and feudal.

The power culture, which is identical to the hierarchy culture in organisations, is dominant in Chilean organisations. Rodriguez and Stewart (2017) further stated that the management of Chilean organisations is known for their dictatorial approach to management. This negative perception of organisations in Chile affects the productivity and performance of construction organisations. Gomez and Rodriguez (2006) also described the Chilean organisational culture as being authoritarian in nature. This is supportive of the hierarchical culture, which is bureaucratic and procedural. In this culture, it is difficult for employees to express their creativity or adopt new approaches in the workplace.

7.8.3 Argentina

Perez-Floriano and Gonzalez (2007) studied the organisational culture of construction companies based in Argentina and Brazil. They identified evidence of a power culture in Argentina. This power culture is very similar to Chile's hierarchical culture, where the management of organisations has a strong hold on decision-making and provides a top-to-bottom communication approach. However, Argentina does not have the authoritarian features of Chilean organisations. In addition, Perez-Floriano and Gonzalez (2007) opined that South American organisational cultures support the inculcation of rules and people in power without much consideration for the employees. This type of culture provides a stringent approach to innovation and in most cases is closed to new ideas and creativity.

7.9 Awareness and acceptance of continuous improvement in the organisational culture

This chapter has reviewed organisational cultures from various countries globally. The intention of this review was to understand how geographical organisational cultures can impact the idea of continuous improvement in the construction industry and, more importantly, continuous cost improvement. The OCAI approach was used to provide a background outline of organisational cultures in various countries (Figure 7.1).

The geographical distribution of the organisational cultures indicates that Africa is predominantly a market culture, whereas instances of adhocracy are found in Germany and Australia. The clan culture is widespread in Asia and North America, while India has a mixture of clan and hierarchy cultures. Nigeria has a mixture of hierarchy and market cultures. Apart from Brazil, South American countries exhibit a hierarchy culture, thus making the inculcation of change in their various organisations difficult. However, wherever adhocracy and clan culture are predominant, there is an opportunity for change, innovation, and continuous improvement.

In the next sections, the impact of these cultures on the penetration of continuous improvement concepts is discussed.

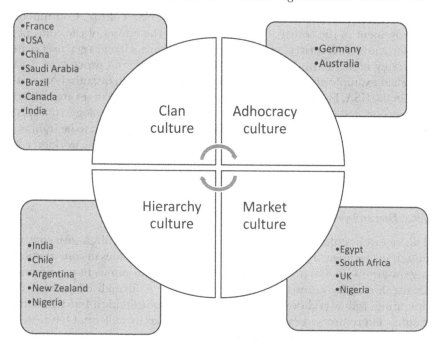

Figure 7.1 Summary of organisational cultures around the world.

Source: Authors.

7.9.1 Clan culture and continuous improvement

The clan culture creates a family-like environment that fosters a positive relationship between the employees and the management of organisations. In the construction industry, the clan culture is essential for getting things done. However, in certain extremes where nepotism and sentiments are evident in the clan culture, it may be very difficult for innovation and change to manifest. An example of the extreme feature of the clan culture is Saudi Arabia. In the extreme clan culture, religious and social cultures permeate organisational culture. This extreme clan culture can make it difficult for continuous improvement practices to penetrate construction organisations. Conversely, continuous improvement is possible in a clan culture when the management decides to invest in training and human resource development schemes. The clan culture provides a basis for internal stability in construction organisations, thereby enshrining cogent aspects of internal consistency, standardisation, and improvement of processes. The features of clan culture can be compared to continuous improvement when there is a focus on client satisfaction.

Client satisfaction in the construction industry is paramount to enhancing competitive advantage and retaining existing customers. Collaboration through modern information management systems such as BIM is easily feasible in construction organisations with clan culture. Continuous improvement in the construction industry with the advent of the concepts of construction Industry 4.0 and the clan culture will be very useful in delivering new training, strategy, mission statements, and processes.

A clear example of positive clan culture can be found in organisations situated in the USA, France, and Brazil. In these countries, employees are trained and encouraged to innovate, and the adoption of new technology, concepts, and ideas is vast. Continuous improvement has been explained to be training within the workforce in the USA. This feature of training in the workforce is common in the clan culture where the management intends to make use of the positive morale to drive organisational growth harmoniously.

7.9.2 Hierarchy culture and continuous improvement

The hierarchy culture has some positive attributes which align with continuous improvement. The standardised structure of activities in construction organisations is comparable with the feature of continuous improvement. In the hierarchy culture, organisations follow a formalised pattern of operating, follow laid down rules, and ensure coordination for delivering results. There are several extremes to the hierarchy culture. One of the extremes is the case of Chile, where organisations are authoritarian and do not permit flexibility and change. The extreme of authoritarian leadership in construction organisations may only create effectiveness and discipline in delivering construction outputs. The attributes of client satisfaction, relationship management, training in the workplace, and innovation are present in the hierarchy culture.

Continuous improvement thrives when there is a sense or philosophy of "How can we do things better?" The Japanese translation of continuous improvement is *kaizen* which implies "change for better". Therefore, the change mantra is the bedrock of continuous improvement. In the least extreme hierarchy culture, change is not easily accepted because the management strategy looks at sustaining the *status quo ante*, which has been effective and efficient in delivering the desired outcomes.

The hierarchy culture may tend to focus on organisational performance rather than the satisfaction of employees and clients. Performance in terms of profit attainment and construction output delivery time, cost, and quality are the basic tenets of a hierarchy culture. Continuous improvement may be difficult to attain in construction organisations with a hierarchy culture because of the dearth of openness to new ideas and innovation.

7.9.3 Market culture and continuous improvement

Like the hierarchy culture where the outcome is more important, the market culture is based on the outcome of processes and activities. The market culture is based on the external economic, socio-political, and regulatory policies which influence organisations. The impact of the market culture is that competition among construction organisations will drive individuals to become more focussed and positively aggressive towards achieving their goals. The market culture is driven by the desire to be more competitive and enjoy prominence. Hence, leadership is important, and employees are viewed are micro-leaders who must endeavour to champion the goals of the organisation.

The market culture may not be entirely open to continuous improvement because of individual extremes and organisational differences. However, in specific extremes, innovation may be viewed as an opportunity to gain a competitive advantage. A clear example of an innovation-driven market culture is the UK. In the UK, the government strategy for BIM adoption ensured that many construction organisations involved in public construction projects made use of BIM. This approach was driven by the external market and influenced by the government; therefore, many construction organisations began to adopt BIM. Continuous improvement under the market culture may be forced and difficult to attain. However, when compared with the hierarchy culture, the market culture can accept continuous improvement concepts.

The key attribute linking the market culture with continuous improvement is competition. Continuous improvement seeks opportunities to make organisations more competitive through waste minimisation, innovative production efficiencies, and constant change. The market culture can respond to constant change; however, this is mainly from external influences such as government policies and directives, inflation, exchange rate, interest rates, and changes in government.

7.9.4 Adhocracy and continuous improvement

The adhocracy culture in organisations places more importance on innovation, creativity, and individual initiative. Very few organisations globally have this approach towards their mission statement and employee engagement. The extreme aspect of an adhocracy culture is the nature of risks which the management permit employees to take and the level of trust existing between employees and upper management. This culture must be controlled effectively; otherwise the intended outcomes may not be attained owing to failed experiments. The adhocracy culture is very common in German and Australian construction organisations. In these countries, construction Industry 4.0 concepts, for instance, were easily adopted, thus serving as an indication that continuous improvement in the construction industry remains feasible under the adhocracy culture in construction organisations.

The long-term agenda of adhocracy culture is to produce new assets by ensuring continuous improvement principles in their organisation. Since individuals in adhocracy culture have creative freedom, this culture makes it easier for continuous improvement to be implemented in their construction organisations. When adhocracy culture is compared with the clan, hierarchy, and market cultures, it is more open to innovation and creativity-driven ideals of continuous improvement.

7.10 Summary

This chapter has reviewed the broader perspectives of organisational culture using the OCAI approach of clan, hierarchy, market, and adhocracy cultures. It is easier for continuous improvement to be adopted in an adhocracy culture. The existence of an adhocracy culture in a construction organisation may be evidence of continuous improvement in the construction industry, which may have not been documented. Additionally, the continuous cost improvement can only be possible in situations where continuous improvement has become a culture in a construction organisation.

References

Abiola-Falemu, J. O., Ogunsemi, D. R. and Oyediran, O. (Eds.). (2010). In *Proceedings of the W117 – Special Track 18th CIB World Building Congress*, May 2010, Salford, UK.

Alao, T. O. and Aina, O. O. (2020). Organisational culture of construction companies in Lagos State, Nigeria. *Organization, Technology and Management in Construction, 12*(1), pp. 2158–2169. doi: 10.2478/otmcj-2020-0012.

Aldraehim, M., Edwards, S., Watson, J., and Chan, T. (2012). Cultural impact on e-service use in Saudi Arabia: The role of nepotism. *International Journal for Infonomics, 5*(3–4), 655–662.

Al-Sedairy, S. T. (2001). A change management model for Saudi construction industry. *International Journal of Project Management, 19*(3), 161–169. doi: https://doi.org/10.1016/S0263-7863(99)00067-8.

Ankrah, N. A. and Proverbs, D. (2005). A framework for measuring construction project performance: Overcoming key challenges of performance measurement. In *Proceedings of the 21st Annual Conference of the Association of Researchers in Construction Management, ARCOM 2005*, 2 September, pp. 959–969.

Calori, R. and Sarnin, P. (1991). Corporate culture and economic performance: A French study. *Organization Studies, 12*(1), 49–74. doi: 10.1177/017084069101200104.

Cameron, K. S. and Quinn, R. E. (1999) *Diagnosing and changing organisational culture: Based on the Competing Values Framework*. Reading, PA: Addison-Wesley.

Cheung, Y., Rowlinson, S., Spathonis, J., Sargent, R., Jones, T., Jefferies, M., and Floriente, G. (2004). Organisational structure, culture and commitment: An Australia public sector case study (CS). In *Proceedings of Clients Driving Innovation International Conference 2004* (pp. 79–89). CRC for Construction Innovation, Queensland, Australia.

Dajani, M. A. Z. and Mohamad, M. S. (2017). Leadership styles, organisational culture and learning organisational capability in education industry: Evidence from Egypt. *International Journal of Business and Social Research, 6*(11), 42. doi: 10.18533/ijbsr.v6i11.1022.

Dastmalchian, A., Lee, S. and Ng, I. (2000). The interplay between organizational and national cultures: A comparison of organizational practices in Canada and South Korea using the competing values framework. *International Journal of Human Resource Management*, *11*(2), 388–412. doi: 10.1080/095851900339927.

Dawson-Shepherd, A. (1997). Communication in organisations operating internationally. *Journal of Communication Management*, *2*(2), 158–166. doi: 10.1108/eb023456.

Dlamini, G. T. (2015). *Organizational Culture in the South African Construction Industry: Effects on Work-Life Balance and Individual Performance* (Doctoral dissertation, Nelson Mandela Metropolitan University), Gqeberha, South Africa.

Durdyev, S. and Mbachu, J. (2011). On-site labour productivity of New Zealand construction industry: Key constraints and improvement measures. *Australasian Journal of Construction Economics and Building*, *11*(3), 18–33. doi: 10.5130/ajceb. v11i3.2120.

El-Nahas, T., Abd-El-Salam, E. M. and Shawky, A. Y. (2013). The impact of leadership behaviour and organisational culture on job satisfaction and its relationship among organisational commitment and turnover intentions. A case study on an Egyptian company. *Journal of Business and Retail Management Research*, *7*(2), 1–31.

Engelen, A., Flatten, T. C., Thalmann, J., and Brettel, M. (2014). The effect of organizational culture on entrepreneurial orientation: A comparison between Germany and Thailand. *Journal of small business management*, *52*(4), 732–752.

Felipe, C. M., Roldán, J. L. and Leal-Rodríguez, A. L. (2017). Impact of organizational culture values on organizational agility. *Sustainability (Switzerland)*, *9*(12), 2354. doi: 10.3390/su9122354.

Frey, R., Boyd, M., Foster, S., Robinson, J., and Gott, M. (2016). What's the diagnosis? Organisational culture and palliative care delivery in residential aged care in New Zealand. *Health & Social Care in the Community*, *24*(4), 450–462.

Gomez, C.F. and Rodriguez, J.K. (2006). Four approximations to Chilean culture: Authoritarianism, legalism, fatalism and compadrazgo. *Asian Journal of Latin American Studies*, 19(3), 43–65.

Gray, J. H., Densten, I. L. and Sarros, J. C. (2003). Size matters: Organisational culture in small, medium, and large Australian organisations. *Journal of Small Business & Entrepreneurship*, *17*(1), 31–46. doi: 10.1080/08276331.2003.10593311.

Harinarain, N. and Bornman, C. (2013). Organisational culture of the South African construction industry. *Acta Structilia*, *20*(1), 22–43. Available at: https:// journals.ufs.ac.za/index.php/as/article/view/132/125.

Hofstede, G. (1984). *Culture's consequences: International differences in work-related values* (Vol. 5). London: Sage.

Jaeger, M., Yu, G. and Adair, D. (2017). Organisational culture of Chinese construction organisations in Kuwait. *Engineering, Construction and Architectural Management*, *24*(6), 1051–1066. doi: 10.1108/ECAM-07-2016-0157.

Liu, A. M. M., Shuibo, Z. and Meiyung, L. (2006). A framework for assessing organisational culture of Chinese construction enterprises. *Engineering, Construction and Architectural Management*, *13*(4), 327–342. doi: 10.1108/09699980610680153.

Maiti, A. and Indhu, B. (2018). Organizational culture and its impact in Indian construction industry – A case study. *International Journal of Civil Engineering and Technology*, *9*, 110–125.

Nazarian, A., Atkinson, P. and Foroudi, P. (2017). Influence of national culture and balanced organizational culture on the hotel industry's performance. *International Journal of Hospitality Management*, *63*, 22–32.

Newman, J. (2007). An organisational change management framework for sustainability. *Greener Management International*, *57*, 65–75. Available at: http://www.jstor.org/stable/greemanainte.57.65.

O'Donnelle, O., and Boyle, R. (2008). *Understanding and managing organisational culture*. Dublin, Ireland: Institute of Public Administration.

Omotayo, T., Awuzie, B., Egbelakin, T., Obi, L., and Ogunnusi, M. (2020). AHP-systems thinking analyses for Kaizen costing implementation in the construction industry. *Buildings*, *10*(12), 230.

Perez-Floriano, L. R. and Gonzalez, J. A. (2007). Risk, safety and culture in Brazil and Argentina: The case of TransInc Corporation. *International Journal of Manpower*, *28*(5), 403–417. doi: 10.1108/01437720710778394.

Rodriguez, J. K. and Stewart, P. (2017). HRM and work practices in Chile: The regulatory power of organisational culture. *Employee Relations*, *39*(3), 378–390. doi: 10.1108/ER-02-2017-0034.

Seibel, W. (1990). Government/third-sector relationship in a comparative perspective: The cases of France and West Germany. *Voluntas: International Journal of Voluntary and Nonprofit Organizations*, *1*(1), 42–60.

Strossmayer, J. J., Matotek, J. and Strossmayer, J. J. (2014). Importance and trends of organizational culture in construction in eastern Croatia. *Ekonomski Vjesnik*, *27*(1), 25–40.

Thomas, R., Marosszeky, M., Karim, K., Davis, S., and McGeorge, D. (2002). The importance of project culture in achieving quality outcomes in construction. *Proceedings IGLC*, *10*, 1–13.

Von Meding, J., Kelly, K., Oyedele, L., and Spillane, J. (2012). Stakeholder management and corporate culture in the UK construction industry. In *RICS COBRA Conference 2012, Las Vegas, Nevada, USA*.

Von Meding, J., McAllister, K., Oyedele, L., and Kelly, K. (2013). A framework for stakeholder management and corporate culture. *Built Environment Project and Asset Management*, *3*(1), 24–41. https://doi.org/10.1108/BEPAM-07-2012-0042

Zhang, S. B. and Liu, A. M. M. (2006). Organisational culture profiles of construction enterprises in China. *Construction Management and Economics*, *24*(8), 817–828. doi: 10.1080/01446190600704604.

8 Continuous Improvement and Cost Overrun in Construction Projects

Temitope Omotayo, Udayangani Kulatunga, and Bankole Awuzie

8.1 Introduction: Classification of construction projects

The scale of construction projects has been a subject of debate among project management researchers. For instance, Santana (1990) divided construction projects into three main categories, namely normal, complex, and singular projects. Santana (1990) explained that singular projects are distinct, longer, and more sporadic. Normal projects are mostly championed by government and multinational corporations with larger investments, coupled with convoluted systems of construction management. The output of normal projects led by larger institutions provides greater economic benefits. Complex projects are usually industrial projects which share many of the attributes of singular projects but are mostly conducted outside townships or residential locations.

In another classification conducted, Safa, Sabet, MacGillivray, Davidson, Kaczmarczyk, Gibson, and Rayside (2015) identified mega, large, complex, and basic projects as five categories of construction projects. The classification of construction projects is not only defined by their financial outlay, but also includes duration, project stakeholders, size of the project, and economic benefit. Therefore, the salience of improving construction processes pervades construction project and cost management domains. Irrespective of the nature of complexity in construction projects, it is important to understand the cost performance of construction projects regarding their complexity and the dynamics involved. Hence, this chapter evaluates the scale of construction projects according to the categorisation of Safa et al. (2015) and their cost performance variables.

8.2 Continuous improvement in basic projects

Basic or normal construction projects are projects which cost less than $10 million and are normally executed within a year (Safa et al., 2015). Examples of basic projects are renovations, extensions, new buildings, earthworks, and road projects. Basic construction projects may be small-scale in size and very few stakeholders will be involved when compared to

DOI: 10.1201/9781003176077-10

other variations of construction projects. Basic construction projects can be classified as minor works. Continuous improvement can be applied in construction projects to eliminate defects with the aid of plan-do-check-act (PDCA) principles (Gonzalez Aleu & Van Aken, 2016). Gonzalez Aleu and Van Aken (2016) further noted that the PDCA principle led to the identification of lessons learned during the construction process. Gieskes and Ten Broeke (2000) observed that continuous improvement was essential for creating learning organisations through construction projects. Hence, the focus should be on improving construction projects processes as a tool for construction organisation improvement.

Furthermore, other aspects of basic construction projects such as design and planning, risk management, stakeholder management, information management, procurement, and contract formation are vital in the construction project life cycle. Rodgers (2017) advocated for the application of continuous improvement in the construction industry in an era of scarce resources. Cost management practices in basic construction projects are confronted by certain specific challenges which may limit the incorporation of continuous improvement. Hence, to understand the underlying determinant of cost escalations in basic construction projects, the next section addresses some essential factors.

8.2.1 Factors impacting on the cost performance of basic construction projects

The causative factors for cost-overruns on basic projects can be categorised as employer, contractor, consultant, construction material, labour, and equipment, contract, contractual relationship, and external factors (Alwi & Hampson, 2003; Rodgers, 2017). Generally, the causation of cost overrun may be divided into design and planning errors, uncertain weather conditions, force majeure, and variations (Rodgers, 2017). In road construction projects Love, Ahiaga-Dagbui, and Irani (2016) observed a disparity of cost overrun from 11% to 106% across various countries. Hence, the benchmark for determining construction cost overrun is any increment in the budgeted cost above 5%.

Table 8.1 summarised the causations of construction overruns in basic construction projects. Some of the aforementioned broad areas of construction of external factors, contractual relationships, labour, and equipment may be similar to the overrun challenges in complex, large, and mega construction projects. However, the main categories under these headings will be unique.

8.2.1.1 Employer-related

The employers' actions may lead to cost overrun in basic construction projects when they fail to finance the project and make payments on time. Additionally, the negative actions of the employer may change the scope

Table 8.1 Causation of cost overrun in basic construction projects

S/N	Category	Causation of cost overrun
1.	Employer	Project finance and payment
		Negative interference from the employer
		Slower decision-making by the employer
		Unrealistic contract duration imposed by the employer
2.	Contractor	Issues with the subcontractor
		Poor site management
		Construction method
		Improper planning
		Mistakes during construction
		Inexperience contactor
3.	Consultant	Preparation and approval of drawings
		Contract management
		Quality assurance and control
4.	Construction material	Quality of construction material
		Shortage of construction material
5.	Labour and equipment	Shortage of labour supply
		Labour productivity
		Availability of equipment
		Failure of equipment
6.	Contract	Changer orders
		Discrepancies in contract documents
7.	Contractual relationship	Disputes
		Project organisation
		Communication issues
8.	External factors	Adverse weather conditions
		Changes in regulations
		Community engagement issues
		Unforeseen ground conditions

Source: Adapted from Alwi and Hampson (2003).

of the project and limit the productivity of the contractor with unrealistic expectations. Delayed decision-making by the employer may also lead to construction delays and eventually, overrun in construction cost. In preventing these challenges, continuous improvement provides a clear approach to timely intervention and proactive planning strategies in contract management and decision-making.

8.2.1.2 Contractor-related

The contractors' inexperience can lead to poor construction site management; adoption of the wrong construction method; mistakes in construction; and improper planning activities. The contractor is responsible for the construction project and the appointment of sub-contractors. Therefore, when subcontractor's mistakes occur, the contractors are deemed responsible for the actions of the sub-contractor. The contractor's experience in interpreting the contract and ability to deliver the project are crucial in ensuring that construction projects are delivered within the set budget. Continuous improvement practices under lean construction comprising of construction methods which engender waste reduction; standardisation; integrated supply chain management; and continuous training of on-site labourers may be adopted by contractors to mitigate the challenges associated with cost-effective project delivery. The only continuous improvement practice which can mitigate the facet relating to inexperience is constant training and development on the part of the contractor. Post-project reviews of completed construction projects may also support contractors' experience development.

8.2.1.3 *Consultant-related*

Consultants involved in design are mainly architects and civil and structural engineers while the cost planning may involve cost engineers or quantity surveyors. Construction managers are responsible for providing the schedules and construction method statements, including the conditions of contract documentation. Errors in construction projects emanate from designs, bills of quantities, schedules, construction method statements, and contracts are linked with the construction cost overruns through variation and change orders. Therefore, issues with contract management and quality assurance depend on the role of consultants and the contract administrator. A continuous improvement practice of post-project reviews of construction contract documentation and value management is essential for improving the accuracy of documents produced by consultants.

8.2.1.4 *Construction materials-related*

Construction material delivery to challenging construction sites, importation and transportation cost are some of the causations of cost overrun when an integrated approach has not been adopted for the supply chain. In some situations, contractors may be faced with the challenge of construction material shortage and poor quality of construction material. Construction material shortage is a major causation of cost overrun. Instances of inflation and fluctuation of prices can also cause cost overruns. The quality of building material has been a factor influencing variation, scope changes, and disputes on construction sites. Total quality management and lean six sigma practices in the construction industry are features of continuous improvement capable of ensuring the right construction material quality is used.

8.2.1.5 *Labour and equipment-related*

The supply of labour on construction sites has become a major determinant of project costs. If there is a shortage, there is a possibility of time overruns which will lead to cost overruns. Similarly, labour productivity affects construction quality and the timeliness of project delivery. Availability or failure of equipment on construction sites affects labour productivity. In basic construction projects, the equipment downtime will impact on the cost of hiring, repair, and depreciation. Integrated teams, early involvement of contractors, and logistics management are essential to eliminating labour and equipment challenges effectually on construction sites.

8.2.1.6 Contract-related

Change orders result from variation owing to errors and omissions from construction activities. Discrepancies in construction contract documents can lead to variations and change orders. Furthermore, improper contract management of basic construction projects may result in disputes when the claims are rejected. Total quality management techniques and PDCA on construction sites can aid the detection of the design errors before they are transferred to the construction process.

8.2.1.7 Contractual relationship-related

The inability of the contractor, employer, or other key construction stakeholders in managing contractual relationship may lead to disputes. Disputes are a major causation of cost overrun in all forms of construction projects. Therefore, improper construction project organisation and communication can be traced to disputes. One of the benefits of continuous improvement in construction project management is effective communication while site meetings and standardisation are some of the guiding principles of continuous improvement capable of reducing the impact of negative contractual relationships on construction projects.

8.2.1.8 External factors

Adverse weather conditions, like heavy rain, blizzard, strong wind, or hailstorms are foreseen risks that may delay the delivery of projects and invariably lead to higher construction costs. External factors are mostly uncontrollable and may result in higher construction costs. Unforeseen ground conditions in road construction projects can also add considerably to the cost of ground investigation and overall construction costs. However, restrictive government regulations and community engagement are two key factors that can limit the cost of construction projects. Nevertheless, the inclusion of insurance, indemnity, and force majeure clauses in construction contracts is a proactive approach to mitigating the impact of external and unforeseen events on construction projects.

8.3 Complex construction projects

Complex construction projects have an average cost ranging from $10 to $100 million, and are executed within one to three years (Safa et al., 2015; Santana, 1990). Examples of complex construction projects are industrial, urban development, and public infrastructure works such as bridges, public event centres, airports, rail projects, harbours, and terminals. A construction project may be categorised as complex when additional planning is required for safety, technological requirements, and increased coordination of logistics and supply chain for material delivery. Santana (1990) opined that complex construction projects, however, share certain features of basic construction projects. This implies that some of the design, planning, risk management, and contract formulation in complex construction projects will be similar to basic construction projects. However, the scale of stakeholder management and construction technology essential for execution is higher in complex construction projects. Landwójtowicz (2018) presented case studies on how continuous improvement has enhanced the performance of projects through the adoption of standardisation of the planning, construction, and reporting processes. Continuous improvement is essential in construction progress and construction method effectiveness monitoring (Landwójtowicz, 2018). Other areas for improvement in complex construction projects were identified by Serpell and Alarcón (1998) including the following:

i   Site layout improvement which leads to a reduction in time taken for transportation and access to construction materials
ii   Decline in the number of on-site personnel
iii   Improvement of executing planning
iv   Training of on-site workers
v   Improved implementation of operational planning
vi   Enhancement of construction methods from masonry to concrete

Continuous improvement of complex construction projects can be implemented in the areas of logistics, on-site management of the construction process and labour, and construction technology. Serpell and Alarcón (1998) further noted that the application of continuous improvement in a complex construction project resulted in a 25% increase in labour productivity, a visible decline in the construction cost, and a 70% decrease in employer complaints. The outcome of implementing continuous improvement in a complex project echoes the results documented by other researchers such as Omotayo et al. (2020) and Vivan, Ortiz, and Paliari (2015). The impact of continuous improvement becomes more relevant in larger projects.

8.3.1 Factors impacting on the cost performance of complex projects

Apart from the factors identified in Tables 8.1 and 8.2, other broader determinants influencing the cost performance of complex construction projects are the following:

i The location of the site
ii Political and planning requirements
iii Procurement and financing
iv Government regulatory requirements

Projects financed through public-private partnerships will necessitate more complex collaborative arrangements, contractual obligations, and capital outlay. The location of the project may also have a greater impact on the cost performance of complex projects. A study of complex highway projects with costs spanning between $10 to $50 million was conducted by Creedy, Skitmore, and Wong (2010). The findings indicated that there was a cost overrun of 20% to 23.2% on these projects. The fundamental factors impacting complex projects are summarised under the headings in Table 8.2.

Table 8.2 is inclusive of all the causations of cost overrun in basic construction projects. Under the complex construction projects, the categorisation of cost overrun causes includes design-related factors, contractor-related factors, construction process-related factors, construction materials-related factors, and external factors.

8.3.1.1 Design

Changes in design and scope during construction may either be minor or major. Irrespective of the nature of changes, the scale of complex projects will be more highly impacted by changes in design and scope.

Table 8.2 Causations of cost overrun in complex projects (inclusive of Table 8.1)

S/N	Category	Causation of cost overrun
1.	Design	Design and scope change
		Deficient documentation
		Specification and design errors
2.	Contractor	Contractor risks
		Errors in construction project management costs
3.	Construction	Constructability
		Services relocation
		Right-of-way costs
		Environment
4.	Construction material	Price escalation
		Latent conditions
		Insufficient investigation

The consequence of such changes, which may also result from deficient documentation and errors in the specification is a significant overrun in cost. Cost escalations are very common in construction projects where design-related information has been transferred into the execution phase.

8.3.1.2 Contractor

Contractors' risk in construction projects pertains to cost allocations for preliminary items of works which are components of their overheads, and relevant unforeseen occurrences resulting from operations on construction sites. Construction project management cost also forms a major part of the overhead costs, which must not be higher than expected. Consequently, risks emanating from contractors' risks and project execution costs may be as a result of errors in estimations. The outcome of errors in estimation leads to construction cost overrun, which may be borne by the contractor, depending on the contractual arrangement.

8.3.1.3 Construction

Constructability or buildability is a design risk that may not be known until the execution phase of a contract. Constructability pertains to the ease of construction or ability to transfer the design ideas into reality on construction sites. Hence, constructability may lead to design changes and additional costs if the design is not feasible for construction. The location of underground services may require additional surveying efforts to make way for the construction process. Right-of-way cost which deals with access to construction sites may be included in the overhead costs. However, when the right-of-way costs and relocation of underground service are not identified early before construction, they potentially escalate the cost. The nature of the environment where construction activities is conducted can create more challenges for the construction and therefore incur additional costs. Hence, the PDCA continuous improvement principle consistently identifies problem areas in construction for the purpose of proactively eliminating cost escalation challenges before they impact the whole construction project.

8.3.1.4 Construction materials

Latent unidentified ground conditions such as groundwater, loose soil conditions, underground boulders, and environmental impacts resulting from the construction processes can contribute to cost overruns. Escalations in the price of construction materials are another major factor that may derail the cost forecasts of construction projects. In unstable economies where the price of construction materials such as cement, timber, and steel, may escalate, it is necessary to create provisions proactively for changes in inflations, exchange rate, and escalations.

8.4 Continuous improvement on large and unique construction projects

Large construction projects are characterised by their high costs which run to over $100 million and are under $1 billion. Large construction projects are executed for more than two years under public works with multiple contractors, specialists, and consultants (Safa et al., 2015). Large construction projects are different from complex construction projects in terms of their financial investment, class of client which is mainly the government, and the number of contractors involved. Complex construction projects may be similar to large projects in terms of the nature of project stakeholders' dynamics. Conversely, large construction projects should not be confused with construction megaprojects because the main differences are not the type of construction projects by the financial investment, duration, and number of stakeholders. The scale of managing large public construction projects is more complicated when compared with complex construction projects. Examples of large construction projects are hospitals, power plants, bridges, and energy projects. The design elements, risk management, contract documentation and formulation, and construction in large construction projects require immense planning, analysis, and execution. In large construction projects, the application of Building Information Modelling (BIM) and machine learning can aid execution and cost management (Rahimian, Seyedzadeh, Oliver, Rodriguez, & Dawood, 2020). Site BIM and mobile computing were applied in the construction of a hospital in Europe (Davies & Harty, 2013). The findings by Davies and Harty (2013) buttressed the need to inculcate the continuous application of innovation in construction site process through information technology and best practices. Similarly, continuous improvement strategies will be efficient in large construction projects in the following areas:

i Quality planning, assurance, and planning
ii Adoption of a capability maturity model for project management development
iii Leadership and community engagement
iv Enhanced communication with all key stakeholders including the community
v Construction information management system through BIM.
vi Smart contracts
vii Adoption of the last planner system

The last planner system is a lean construction principle similar to a continuous improvement practice of integrating ideas and managing information from all stakeholders thereby aiding the compression of activity duration, production flow enhancements, waste reduction, and building effective collaboration (Cortés, Herrera, Muñoz–La Rivera, & de Matos, 2017).

The last planner system can be an effective continuous improvement tool in large and unique construction projects where the number of stakeholders cannot be managed with the principles of basic construction projects. Furthermore, the application of mobile project management and smart contracts are innovative practices akin to continuous improvement capable of mitigating collaboration challenges.

8.4.1 Factors impacting on the cost performance of large and unique projects

The factors affecting the cost performance of large and unique construction projects can be categorised according to the following:

   i Procurement method (mostly public–private partnership)
  ii Contractual arrangement
 iii Complexity of project management and organisation
  iv Project financing
   v Economic impact of the project
  vi Environmental conditions and weather
 vii Nature of construction material
viii Strategic value
  ix Duration

Adam, Josephson, and Lindahl (2017) identified and reviewed the causations of cost overruns in large construction projects and categorised the leading factors under the scope of political, economic, technical, and psychological factors. The psychological factors can be associated with optimism or bias, the strategic value of the construction project and the overall expected impact it will deliver. Consequently, the components of cost overruns in large-scale construction projects may be categorised under the structure of Table 8.1, as presented in Table 8.3. Studies conducted by Adam, Josephson, and Lindahl (2017) and Le-Hoai, Lee, and Lee (2008) were used to compile the contents of Table 8.3.

8.4.1.1 Employer/funder

In addition to the causations of cost overrun in Tables 8.1 and 8.2, the unique causations of cost overrun in large and unique construction projects are principally prejudiced by financial difficulties. In public–private partnership projects which compel the private funder to contribute regular finances to the project, there is a possibility of delayed payment for works completed, which can negatively impact on the project delivery cost. Consequently, cost overrun resulting from financial difficulties and delayed payment are significant challenges facing large and unique projects, especially those funded by the joint effort of both public and

Table 8.3 Causation of cost overrun in large and unique construction projects (this table is a continuation, inclusive of Tables 8.1 and 8.2)

S/N	Category	Causation of cost overrun
1.	Employer/funder	Financial difficulties
		Delayed payments for completed works
2.	Contractor	Financial difficulties of the contractor
3.	Construction project management	Ineffective site management and supervision
		Adoption of obsolete construction methods
		Additional works
		Slow inspection of completed works
		Futile monitoring and controls
4.	Contractual relationship	Unproductive communication between contractor and subcontractor
		Dishonesty and deception
		Optimism bias

private employers. The best continuous improvement strategy which may mitigate this cost overrun causation is the development of funding alternatives and the inclusion of contractual provisions for suspension of construction activities on-site pending the resolution of funding challenges. Moreover, equity funders may be invited into the contractual arrangement to meet the needs of the large project.

8.4.1.2 Contractor

When contractors have financial difficulties, the large construction project may experience cost overruns resulting from delays. This is why large and unique projects have multiple contractors handling various segments of the projects. Multiple contractors are different from subcontractors. In large and unique projects, segmentation of the project into work packages is an attribute of continuous improvement in project delivery.

8.4.1.3 Construction project management

There are a number of causations of cost overrun under the niche of construction project management. A major feature is ineffective site management and supervision. This can be linked to the activities of the contractors and project management planning. BIM and mobile computing have been suggested as major strategies for managing construction sites. Site monitoring and control activities may not produce limited cost control measures when the scale of the project is too large. Accordingly, construction methods in large and unique projects must be consistent with the ideals of producing learning projects whereby site meeting and post-project reviews drive continuous improvement in processes and cost management.

8.4.1.4 Contractual relationship

Ineffective management of contractual relationships among all parties, including third parties, may escalate the cost of construction projects. This may begin with optimism bias in managing the parties based on a prior positive outcome of construction projects. Optimism bias can also be linked to dishonesty and deception in reporting the outcomes of monitoring activities from the perspective of the subcontractor or other contractors working on the construction project. Equally, unproductive communication between contractors and subcontractors can be a major reason for the incidence of cost overruns at a very early stage. Continuous process improvement using forecasting and elimination of waste on construction sites, coupled with the timely meeting, reporting, progress tracking, and communications, provides a stronger approach to ensuring the optimism bias is checked and trust is built.

8.5 Continuous improvement on megaprojects

Megaprojects are mostly funded by government and national institutions with capital investments exceeding $1 billion and are executed over a time scale of more than three (3) years (Santana, 1990; Safa et al., 2015). Megaprojects have a greater impact on the social, economic, political, and ecological variances of a country or region where they are located. Safa et al. (2015) described megaprojects as projects which require extensive planning activities. According to Söderlund, Sankaran, and Biesenthal (2017), megaprojects are large complex projects which cost over $1 billion. Söderlund, Sankaran, and Biesenthal (2017) opined that megaprojects have been criticised for the following:

   i Not meeting the expectations of the sponsors.
  ii Not providing the right investment returns for stakeholders.
 iii Not providing strategic value to the investors.
 iv Having no positive impact on the local population and environment.

Conversely, megaprojects are important infrastructural endeavours in which governments must engage to meet the developmental needs of their region or country. Examples of megaprojects around the world are the High-Speed 2 (HS2) rail project in the UK, the Olkiluoto 3 Nuclear Power Plant project in Finland, the Benban Solar Park in Egypt, and the Manhattan Project in the USA, just to mention a few. The aforementioned megaprojects all have a cost which exceeded $1 billion and were executed over a period of three years. One major feature of a megaproject is the uncertainty and evolving requirements. Therefore, continuous improvement may be very useful in megaprojects where it is necessary to create learning projects in megaprojects. The learning project strategy consists of micro post-project reviews at every completion stage. Furthermore, a PDCA champion may be appointed to monitor all the essential aspects of megaprojects as identified by Kardes, Ozturk, Cavusgil, and Cavusgil (2013).

8.5.1 Factors impacting on the cost performance of mega projects

Flyvbjerg (2005) described megaprojects as a political construct of a Machiavellian paradigm which results in cost underestimations, overstated revenues, undervalued environmental impact, and overvalued economic development effects. In the litany of challenges facing megaprojects delivery, cost overrun has been identified as a dominant force (Flyvbjerg, 2005; Cantarelli et al., 2013; Olaniran, Love, Edwards, Olatunji, & Matthews, 2017). A documentary list of principal causations of cost overrun in construction projects as summarised in the categories in Table 8.4 is also supported by the contents of Tables 8.1–8.3, as noted previously. It is difficult to generalise the causations of cost overrun without associating it with a construction project, the geographical location of the project, and benchmarks. Notwithstanding, the collage of cost overrun causations can be categorised under major headings as designated in Tables 8.1–8.3. Flyvbjerg (2007) categorised the causations of cost overrun in megaprojects such as bridges, rail, and road projects into technical, psychological, and political-economic factors. The three main categories encapsulate the tenets of megaproject delivery. In expatiating these categories, Table 8.4 itemises each causation under the three categories as follows.

8.5.1.1 Technical

Cost forecasting and estimation are essential aspects of successful megaprojects. Nevertheless, megaprojects have suffered from cost overruns resulting from inadequate data and models for cost forecasting. Consequently,

Table 8.4 Causation of cost overrun in mega construction projects (this table is a continuation Tables 8.1–8.3)

S/N	Category	Causation of cost overrun
1.	Technical	Inadequate data and models for cost forecasting
		Inexperience of cost forecasters
		Honest mistakes and technical errors in cost forecasting
		Unsuitable or unavailable construction technology
		Technical know-how
		Complexity management
2.	Psychological	Planning fallacy
		Cost forecasting on delusional optimism
3.	Political-economic	Political influence on cost forecasts
		Political pressure
		Economic vagaries
		Deliberate lying in financial reporting

there is a need to adopt artificial neural network and machine learning approaches in cost forecasting. This depends on the experience and technical know-how of cost forecasters and their ability to make use of existing cost data. In megaprojects, honest mistakes may be very costly because of the immense effect they will have on other components of the project. Complexity management is another technical concern for construction project managers. In complexity management, dynamic simulations of future outcomes of the mega construction project, including the cost of construction, are required to ascertain cost and delivery projections.

8.5.1.2 *Psychological*

There is a planning fallacy in many construction projects, which is borne out of optimism bias. Hence, cost forecasts may be based on a fallacy that all will turn out well, as was the case with another previous project. However, no two construction projects are the same. The socio-economic and political situations in the future of an ongoing construction project may thwart the laid-out plans. In mitigating this psychological challenge, continuous improvement thinking and philosophy are guiding principles for the unique nature of every construction project.

8.5.1.3 *Political-economic*

Political pressure to deliver megaprojects within the set budget and schedule may be viewed as a demotivating factor in the construction process. Political and economic influence on the cost forecasting process may also lead to cost overrun when the cost forecasters are under pressure to lie or report favourable financial performance of the project to meet the expectations of political leaders. In the long run, the final cost-value ratio of such a megaproject will be inconsistent with the reality of the construction project and the final outcome will be a cost overrun. Economic vagaries such as rising interest rates, inflation, exchange rates, and fluctuations of construction material prices all have a significant impact on the cost performance of megaprojects. Continuous improvement is difficult to adopt when political-economic challenges influence construction projects. However, there is a chance of including a continuous improvement philosophy in the government parastatals and regulatory bodies, rather than wait for this philosophy to weather any existing political-economic influences on construction megaprojects.

8.6 Categories of stakeholders in construction projects

Aapaoja and Haapasalo (2014) categorised stakeholders in the construction industry according to the level of importance as the primary core team, key supporting participants, tertiary stakeholders, and extended

Table 8.5 Category of stakeholders

NR	Stakeholder category	Stakeholders
1.	Primary team members	Client
		End-user
		Main contractor
		Main designer
2.	Key supporting participants	All other designers
		Specialist contractors
3.	Tertiary stakeholders	Public authorities
		Local government council
4.	Extended stakeholders	Subcontractors
		Material suppliers
		Community

Source: Adapted from Aapaoja and Haapasalo (2014).

stakeholders. A brief breakdown on the constituents of the stakeholders are itemised in Table 8.5.

The four distinct groups of stakeholders have all key participants in all forms of construction projects. Specialist contractors are considered as key supporting stakeholders in the construction industry because they are essentially involved in the delivery of utilities such as gas, electricity, Internet supply, and mechanical and plumbing infrastructure. All other designers, including the cost planning and construction management team, are also considered as key supporting participants. The local government council and statutory authorities are tertiary stakeholders involved in construction projects. The extended stakeholders are the local community, neighbours, material suppliers, and subcontractors. The primary team members involved in construction projects begin with the client, designers, and end-users.

8.7 Summary: Project scales and continuous improvement

Kardes et al. (2013) described the features of megaprojects by juxtaposing the perspectives of size, time, cost, the nature of the schedule and budget, team composition, contracts, customer support system, project requirements, political implications, communication, stakeholder management, change impact, commercial change, level of risk, external constraints, integration, technology, and IT complexity. Table 8.6 summarised the structure of basic, complex, large, and mega construction projects according to the studies of Kardes et al. (2013) and Haas (2009).

All continuous cost improvement strategies such as the last planner system, continuous improvement and lean thinking, waste minimisation, lean six sigma for quality management, PDCA, post-project reviews, site meetings, learning projects, and adoption of emerging innovation and technology

Table 8.6 Summary of construction project scales

No		Basic projects	Complex projects	Large projects	Megaprojects
1.	Cost	<$10 million	>$10–$100 million	>$100 million	>$1 billion
2.	Time	<1 year	>1–3 years	>2–3 years	>3 years
3.	Team composition	Internal, worked together before	Internal and external, worked together before	Internal and external, have not worked together before	Complex structure of varying competencies and performance records
4.	Contracts	Straightforward	Straightforward	Complex	Extremely complex
5.	Project requirements	Understood, straightforward	Understood, unstable	Poorly understood, volatile	Uncertain, evolving
6.	Political implications	None	Minor	Major, impact core mission	Impact core mission of multiple organisations, states, countries
7.	Communication	Straightforward	Challenging	Complex	Arduous
8.	Stakeholder management	Straightforward	2–3 stakeholder groups	Multiple stakeholder groups with conflicting expectations	Multiple organisations, states, countries, regulatory group
9.	Change impact	A single business unit, one familiar business process, and one IT system	2–3 familiar business units, processes, and IT systems	Enterprise, shifts or transforms many business processes and IT systems	Multiple organisations, states, countries; transformative new venture
10.	Level of risk	Low	Moderate	High	Very high
11.	External constraints	No external influences	Some external factors	Key objectives depend on external factors	Project success depends largely on external organisations, states, countries, regulators
12.	Integration	No integration issues	Challenging integration effort	Significant integration required	Unprecedented integration effort
13.	Technology	Proven and well-understood	Proven but new to the organisation	Immature, complex, and provided by outside vendors	Ground-breaking innovation and unprecedented engineering
14.	IT complexity	Application development and legacy integration easily understood	Application development and legacy integration largely understood	Application development and legacy integration poorly understood	Multiple "systems of systems" to be developed and integrated

Source: Adapted from Haas (2009).

can cut across the different aspects of the scales articulated in Table 8.5. Every project is unique and various project scales will require unique cost reduction and maintenance strategies even though there may be some commonalities.

References

Aapaoja, A. and Haapasalo, H. (2014). A framework for stakeholder identification and classification in construction projects. *Open Journal of Business and Management*, 02(01), pp. 43–55. doi: 10.4236/ojbm.2014.21007

Adam, A., Josephson, P. E. B. and Lindahl, G. (2017). Aggregation of factors causing cost overruns and time delays in large public construction projects: Trends and implications. *Engineering, Construction and Architectural Management*, 24(3), pp. 393–406. doi: 10.1108/ECAM-09-2015-0135

Alwi, S. and Hampson, K. (2003). Identifying the importance causes of delays in building construction projects. In *Proceedings of the 9th East Asia – Pacific Conference on Structural Engineering and Construction*, Bali, Indonesia, pp. 1–6.

Cantarelli, C. C., Flyvbjerg, B., Molin, E. J. and Van Wee, B. (2013). Cost overruns in large-scale transportation infrastructure projects: Explanations and their theoretical embeddedness. arXiv preprint arXiv:1307.2176

Cortés, M. J., Herrera, R. F., Muñoz–La Rivera, F. C. and de Matos, B. E. (2017). Key requirements of an IT tool based on last planner system Principales requerimientos de una herramienta TI basada en last planner system.

Creedy, G. D., Skitmore, M. and Wong, J. K. W. (2010). Evaluation of risk factors leading to cost overrun in delivery of highway construction projects. *Journal of Construction Engineering and Management*, 136(5), pp. 528–537. doi: 10.1061/(asce)co.1943-7862.0000160

Davies, R. and Harty, C. (2013). Implementing "Site BIM": A case study of ICT innovation on a large hospital project. *Automation in Construction*, 30, pp. 15–24. doi: https://doi.org/10.1016/j.autcon.2012.11.024

Flyvbjerg, B. (2005). Machiavellian megaprojects. *Antipode*, 37(1), pp. 18–22. doi: 10.1111/j.0066-4812.2005.00471.x

Flyvbjerg, B. (2007). Truth and lies about megaprojects. *Inaugural Speech for Professorship and Chair at Faculty of Technology, Policy, and Management*, Delft University of Technology.

Gieskes, J. F. B. and Ten Broeke, A. M. (2000). Infrastructure under construction: Continuous improvement and learning in projects. *Integrated Manufacturing Systems*, 11(3), pp. 188–198. doi: 10.1108/09576060010320425

Gonzalez Aleu, F. and Van Aken, E. M. (2016). Systematic literature review of critical success factors for continuous improvement projects. *International Journal of Lean Six Sigma*, 7(3), pp. 214–232. doi: 10.1108/IJLSS-06-2015-0025

Haas, K. (2009). *Planting the seeds to grow a complex project management practice.* Retrieved from http://www.kathleenhass.com/WhitePapers.htm

Kardes, I., Ozturk, A., Cavusgil, S. T. and Cavusgil, E. (2013). Managing global megaprojects: Complexity and risk management. *International Business Review*, 22(6), pp. 905–917.

Landwójtowicz, A. K. (2018). Relationship between continuous improvement practices and project management approach. In *Proceedings of the CBU International Conference*, 6, pp. 271–276. doi: 10.12955/cbup.v6.1168

Le-Hoai, L., Lee, Y. D. and Lee, J. Y. (2008). Delay and cost overruns in Vietnam large construction projects: A comparison with other selected countries. *KSCE Journal of Civil Engineering*, 12(6), pp. 367–377. doi: 10.1007/s12205-008-0367-7

Love, P. E. D., Ahiaga-Dagbui, D. D. and Irani, Z. (2016). Cost overruns in transportation infrastructure projects: Sowing the seeds for a probabilistic theory of causation. *Transportation Research Part A: Policy and Practice*, 92, pp. 184–194. doi: https://doi.org/10.1016/j.tra.2016.08.007

Olaniran, O. J., Love, P. E. D., Edwards, D. J., Olatunji, O. and Matthews, J. (2017). Chaos theory: Implications for cost overrun research in hydrocarbon megaprojects. *Journal of Construction Engineering and Management*, 143(2), p. 05016020.

Omotayo, T., Awuzie, B., Egbelakin, T., Obi, L. and Ogunnusi, M. (2020). AHP-systems thinking analyses for Kaizen costing implementation in the construction industry. *Buildings*, 10(12), p. 230.

Rahimian, F. P., Seyedzadeh, S., Oliver, S., Rodriguez, S. and Dawood, N. (2020). On-demand monitoring of construction projects through a game-like hybrid application of BIM and machine learning. *Automation in Construction*, 110, p. 103012.

Rodgers, S. (2017). Continuous improvement? *SAMPE Journal*, 53(3), p. 44.

Safa, M., Sabet, A., MacGillivray, S., Davidson, M., Kaczmarczyk, K., Gibson, G. E. and Rayside, D. (2015). Classification of construction projects. *International Journal of Civil, Environmental, Structural, Construction and Architectural Engineering*, 9(6), pp. 721–729.

Santana, G. (1990). Classification of construction projects by scales of complexity. *International Journal of Project Management*, 8(2), pp. 102–104. doi: 10.1016/0263-7863(90)90044-C

Serpell, A. and Alarcón, L. F. (1998). Construction process improvement methodology for construction projects. *International Journal of Project Management*, 16(4), pp. 215–221. doi: https://doi.org/10.1016/S0263-7863(97)00052-5

Söderlund, J., Sankaran, S. and Biesenthal, C. (2017). The past and present of megaprojects. *Project Management Journal*, 48(6), pp. 5–16. doi: 10.1177/875697281704800602

Vivan, A. L., Ortiz, F. A. and Paliari, J. (2015). Model for kaizen project development for the construction industry. *Gestão & Produção*, 23, pp. 333–349.

9 Latent Continuous Cost Improvement Strategies in the Construction Industry

Temauthitope Omotayo, Udayangani Kulatunga, and Bankole Awuzie

9.1 Introduction

Continuous cost improvement is a new term in the construction industry and it pertains to the incremental reduction of production cost through the construction phase of projects. Kaizen costing is another nomenclature used to denote continuous improvement in construction (Omotayo, Awuzie, Egbelakin, Obi, & Ogunnusi, 2020; Omotayo, Kulatunga, & Bjeirmi, 2018). Kaizen or continuous improvement of construction cost is usually conducted in the form of cost reduction strategies in the planning and construction stages. Case studies of various strategies adopted to deliver continuous cost improvement are presented in this chapter. Continuous cost improvement strategies from Africa, Asia, Europe, America, and Australia are similar in terms of their consistent methodology to drive down the cost of construction during the cost planning process. The following strategies are mainly applicable in the execution phase of construction. The strategies applicable in the pre-construction phase are also identified in the discussion. Examples of continuous cost improvement strategies from various countries are further presented in each section.

9.2 Continuous cost improvement strategies for the basic construction of a basic construction project in Nigeria

The case of a university building construction in Nigeria denotes a basic construction project where the causations of cost overrun are very similar to Table 8.2 in Chapter 8. In developing countries such as Nigeria, the commonalities in cost overrun causation, according to Dada and Jagboro (2007), are within the spheres of project finance and payments, issues with the subcontractor, poor site management, mistakes during construction, contract management, quality assurance and control, shortage of material supply, labour productivity issues; and unforeseen ground conditions. Construction cost overruns are the main challenge facing project delivery. In the continuous cost improvement of basic construction projects, the challenges can be met with the strategy of engaging with suppliers and

DOI: 10.1201/9781003176077-11

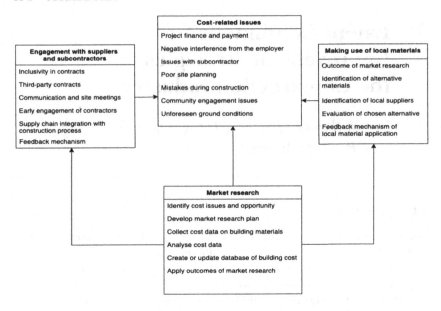

Figure 9.1 Continuous cost improvement strategies in meeting cost-related issues of basic construction projects.

Source: Authors.

subcontractors, making use of local materials, and conducting market research.

The aforementioned strategies are explained in Sections 9.2.1–9.2.3 and Figure 9.1 as drawn from some cost-related issues in Table 8.2. Figure 9.1 identifies cost-related issues of basic construction projects which are in the form of minor renovations and basic construction projects. The indicated cost-related issues in Figure 9.1 are similar to the cost overruns experienced in the delivery of a university building project in Nigeria. These cost-related issues are project finance and payment, negative interference from the employer, issues with the subcontractor, poor site planning, mistakes during construction, community engagement issues, and unforeseen ground conditions.

An explanation of the processes contained in strategically mitigating emerging cost-related issues is based on latent continuous cost improvement strategies which have been adopted by construction cost planners globally. Similarly, the strategies in Sections 9.2.1–9.2.3 are applicable in the pre-contract and post-contract stages of construction projects.

9.2.1 Market research

Market research is a pre-contract continuous cost improvement strategy that is mostly applicable in developing countries in Africa, Asia, and

South America where a construction cost database does not exist. However, market research may be conducted in the post-contract phase of construction in certain circumstances requiring change orders or the emergence of new high-quality alternatives. The process of conducting market research for building materials begins with the identification of cost issues and opportunities. This initial activity is more like a target-costing approach whereby the cost alternatives support the target for each elemental cost (Kern & Formoso, 2006; Pennanen, Ballard, & Haahtela, 2011). In the pre-contract phase of construction, cost issues may be identified with the aid of risk identification, lessons learned from previous projects, and existing cost breakdown in the bills of quantities.

Construction market research must be based on a plan aimed at collecting data on building materials. The sources of cost data may be physical visits to building suppliers and manufacturers and online sources or brochures. The cost data is analysed by a comparative method and a database of building material cost can be created within the organisation. In the UK, the Building Cost Information Service (BCIS) is a common subscription source of updated schedule or rates of building materials and a cost analysis database for a rebased comparative analysis of building construction prices and materials. In developing countries such as Nigeria, where there is no online database of building material prices, construction material price books similar to Spon's Architects' and Builders' Price Book are used. In the process of applying the outcomes of market research, cost targets must be produced to meet the challenge of project financing and payments. Project financing and payments are major causations of cost overrun in building construction. The market research strategy is a continuous improvement strategy because cost data are updated either quarterly or monthly to meet the needs of the construction project and future projects. Market research can be connected with the continuous improvement strategy of making use of local materials to moderate the impact of cost overrun on building construction projects.

9.2.2 *Making use of local materials*

Making use of local materials as a continuous improvement strategy may be applicable in the pre- or post-contract stage of construction. Building construction material scarcity is a hidden causation of delays and cost overrun in construction projects. Ioannidou, Meylan, Sonnemann, and Habert (2017) discussed the scarcity of gravel in certain regions of the world and the need for policymakers to consider local alternatives. Mancini, De Camillis, and Pennington (2013) produced a framework for securing the sources of construction materials. A circular economy approach has been suggested as a major approach to meeting the needs of the construction industry's raw materials. Adopting circular economy practices and making use of local materials are strategic approaches to the continuous improvement of construction cost management.

In the process of making use of local materials, the outcome of the market research must be accepted in the identification of local construction materials such as laterite in Nigeria, which may stand in as a substitute for sharp sand and cement. In the process of identifying local materials, local suppliers may be included in the strategy to engage the community in the construction process. Community engagement issues are a cost overrun challenge in varying scales of construction projects globally. A feedback mechanism in the strategy of making use of local materials from indigenous suppliers will also mitigate any emerging issues with the subcontractor. Consequently, the feedback mechanism on the identification and application of local building materials as alternatives for planned materials will create a continuous process of updating the market research database and further improvements on the quality of local construction materials. The implication of applying local building material alternatives is an engagement with the suppliers and subcontractors.

9.2.3 Engagement with suppliers and subcontractors

Suppliers and subcontractor engagement is central to the success of any project, especially when the vicarious liability has been transferred. Therefore, the first step in engaging subcontractors must be their inclusion in the contract through the third-party right such as collateral warranties. Collateral warranties are third-party contracts created to ensure that the subcontractor is part of an existing contract (Payling, 1992). The integration of the construction supply chain in the construction process has been discussed by Magill, Jafarifar, Watson, and Omotayo (2020) as a process of improving productivity and engaging the suppliers, including the subcontractors. Communication and site meetings must therefore include these parties. Collateral warranties, communication, and site meetings involving the suppliers and contractors will invariably continue to reduce the challenges of site management, mistakes during construction, negative interference by the employers, and the resolution of cost overrun causation such as unforeseen ground conditions. Depending on the scale of a construction project, it is essential to include the subcontractors and suppliers in a contract, site meetings and all essential communication. Incidents of cost overrun can be managed effectively through this strategy. On a larger scale of construction, specific cost issues require specific continuous cost improvement strategies.

9.3 Continuous cost improvement strategies of a complex construction project in Sri Lanka

In the case of a road construction project which is an example of a complex construction project in Sri Lanka as studied by Wijekoon and Attanayake (2012), the leading causations of cost-related issues were identified as

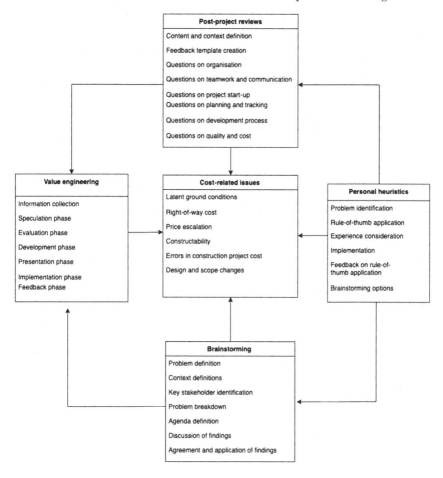

Figure 9.2 Continuous cost improvement strategies in meeting cost-related issues of complex construction projects.

Source: Authors.

constructability, price escalation, errors in designs that were transferred to the estimation process, and latent ground conditions. Likewise, from Table 8.3 in Chapter 8, similar cost-related issues were identified as obtainable in Figure 9.2. In road construction projects' right-of-way, which consists of access to the sites, could be a major issue when underground rock boulders and existing rocks in the construction landscape may lead to a 30% increment in the budgeted cost. Sri Lanka is a developing country and complex construction activities such as roads provide an addition to the much-needed infrastructure in the country. Escalations in the cost of delivering road construction projects in Sri Lanka can further contribute to debates and project financing woes in the country.

Complex construction projects such as road and bridge construction have hidden complexities that are capable of escalating the final construction cost. The application of continuous cost improvement strategies such as brainstorming and post-project reviews, which are driven by personal heuristics, can be merged with value engineering in the execution phase to alleviate the impact of cost-related issues as identified in Figure 9.2.

Individually, the contents of the four strategies indicated in Figure 9.2 are discussed in Sections 9.3.1–9.3.4.

9.3.1 Personal heuristics

Individual heuristics play a vital role in decision making and problem solving in the application of continuous cost improvement. Goodwin, Wright, and Phillips (2004) defined heuristics as cognitive shortcuts that an individual employs to arrive at quick decisions.

Personal heuristics are based on individual experiences, which are applicable in decision-making processes (Sprinkle, 2018). The steps taken to applying personal heuristics in construction decision-making, as discussed by Moyane (2018) and Sprinkle (2018), relate to the identification of a problem to which the rule-of-thumb should be applied and the consideration of personal experience in making decisions. For instance, in a road construction project, there are chances of experiencing latent ground conditions and right-of-way issues which will contribute to additional unexpected costs. The rule-of-thumb application from the personal heuristic strategy from the perspective of the contractor will consider the nature of the contract and what the contract clauses say about unforeseen ground conditions. In the Joint Contracts Tribunal (JCT) contract suit, financial compensations are considered for relevant matters, while time loss can be mitigated with the relevant event clauses. In this example, the rule of thumb, which is based on experience, will not necessarily work alone without reference to existing contract documentation. In the event of the constructability of a road design, the use of personal heuristics will depend on the experience of the contractor. Constructability issues that will lead to higher construction cost may be mitigated by combining personal heuristics with brainstorming or post-project review strategies. The personal heuristics of a continuous improvement expert are spirited in encouraging problem identification, solution derivation, and implementation. Therefore, construction cost in constructability difficulties may be reduced by identifying easier ways of implementing design ideas through the rule-of-thumb approach. Hence, personal heuristics will continually improve based on the experience and influence of other key stakeholders in a project. As a continuous improvement strategy, personal heuristics is subject to biases. These biases may be emotional, environmental, social, economic, political, technological, or other individual idiosyncrasies. Therefore, this strategy must be with other strategies, as illustrated in Figure 9.2. The brainstorming strategy for continuous improvement provides an opportunity to enhance the output of personal heuristics. Brainstorming is another continuous improvement strategy that depends on the rule of thumb of many key stakeholders in a construction project.

9.3.2 Brainstorming

Brainstorming is a decision-making strategy that is based on a collection of information from all employees in an organisation. Seaker and Waller (1996) opined that brainstorming sessions potentially fast-track the outcomes of continuous improvement in organisations. In construction projects where continuous improvement is being implemented, all stakeholders can make use of a brainstorming strategy to identify opportunities for cost reduction (Yazdani & Tavakkoli-Moghaddam, 2012). Brainstorming can be applied in the pre- and post-contract construction phases. In the pre-contract phase, brainstorming can be linked with value planning, target costing, and briefing. During the post-contract phase, value engineering, communication and teamwork, site meeting, and post-project reviews can be combined with brainstorming.

In a road construction project in Colombo, Sri Lanka, the quantity surveyor involved identified the application of brainstorming as a strategy for reducing construction cost during the cost planning phases with the inclusion of value engineering techniques during the construction stage. During the construction phase, a brainstorming session involves the subcontractors and suppliers. A site meeting is held on the progress of construction, material delivery, the outcome of the interim valuation, and cash flow projects. Alternatives are generated as options to keep the overhead costs at a minimum, study the outcomes of projected costs and proffer solutions to plausible occurrences of cost overruns. The brainstorming approach has been an effective instrument in site meetings, value engineering, the application of personal heuristics, market research, constructability assessment, and the overall basis of continuous cost improvement. Khan, Kaviani, Galli, and Ishtiaq (2019) maintained that brainstorming is a problem-solving technique in the continuous improvement process. The PDCA technique largely depends on the activities of all employees on the project to identify problem areas. Likewise, the last planner system, which is a lean construction and a technique for continuous improvement in terms of project planning, also depends on these techniques for brainstorming.

The first phase of a brainstorming session starts with problem definition. In terms of the context of this chapter, price escalation resulting from errors in project cost estimation (as recognised in Figure 9.2) is a major cost problem which brainstorming sessions may resolve. Key stakeholders such as the construction project manager, the quantity surveyor or cost planner, the contract administrator or architect, subcontractors, and the main contractor, along with the suppliers and the employer, can form a focus group to brainstorm available solutions. A breakdown of the problems identified will lead to the application of personal heuristics in offering solutions. The implications of recommended solutions will be weighed against other options and a final agreed solution will be implemented. Brainstorming as a continuous improvement strategy provides an opportunity for effective communication and teamwork in construction projects. Brainstorming sessions also create a cohesive arrangement in the construction process. Value engineering depends on the brainstorming strategy: its individual processes are discussed in the next section.

9.3.3 Value engineering

Value engineering is a value management system method of driving down construction cost during the execution phase of a project (Cheah & Ting, 2005; Khodeir & El Ghandour, 2019). This strategy identifies alternative construction materials at a lower cost and a higher quality during the construction stage (See Chapter 3, Section 3.3). Value engineering, a continuous improvement strategy, uses the strategy of brainstorming to collect information on a problem. In the information collection phase, the scale of the problem is identified and all available contexts are drawn for the speculation phase. As an example from Figure 9.2, value engineering can be applied in the design and scope changes section through brainstorming on why there should be a design and scope change. This deliberation can adopt the key stakeholders identified in the brainstorming session of 9.3.2.

The speculation phase of value engineering must indicate the function of the design and scope changes with considerations of implications on the construction cost. The creative phase adopts the content of the brainstorming session. The best value and applicability of all options for changing the design and scope of the project can be discussed in the evaluation phase. The developmental phase considers the options for improvement of the design and scope without having much negative impact on the cost of construction. The presentation phase is where the outcome of the value engineering brainstorming session is reviewed by all the main stakeholders, in this case, the employer, contractor and contract administrator for their final approvals. The implementation and feedback phase leads to a continual evaluation of the best possible options to effect changes in the design and scope to provide optimal value at the lowest cost, subsequently leading to a continual improvement of the construction cost along with the design and scope of the project. Post-project reviews may be essential in delivering the value engineering strategy for continuous cost improvement because they may elucidate previous project options.

9.3.4 Post-project reviews

Post-project reviews, as discussed previously in Section 4.8 of Chapter 4, are a construction cost management technique for evaluating project performance. Post-project reviews in the context of continuous improvement are a construction feedback mechanism conducted through site meetings and end-of-project review meetings (Carrillo, Harding, & Choudhary, 2011; Omotayo & Kulatunga, 2017). Post-project reviews can be carried out with a feedback template which will be inclusive of the questions raised in Figure 9.2. Just as every other continuous improvement strategy has identified and defined the context of a problem, a post-project review must have a review context, and, in this instance, an entire road or complex construction project may be reviewed. The activity may be conducted by the construction project management working on behalf of the contractor or the employer's architect in charge of the project. A continuous improvement champion may also lead the project review process.

The first question in reviewing completed projects concerns project cost planning, contract documentation, and management. By means of this question the ease of making use of the contract documents such as drawings, bills of quantities, a Gantt chart, a construction method statement, and conditions of contract can be evaluated.

The balance of resources allocation in important aspects of the project can be considered for feedback. The delivery of the plans on-site can be considered in terms of what went well and what could have been improved. A rating system on the scale of 1–5 or 1–10, similar to the Likert scale, can be used to determine the performance of the project organisations. This similar template for extracting feedback from all stakeholders in the project may pose further questions in the line of teamwork and communication.

The quality of construction project team communication can be reviewed from an internal perspective and compared with external communication with subcontractors, suppliers, the community, and other external stakeholders. Accessibility to important decision-makers in the project and information exchange in all phases of the project must be reviewed. The construction project which starts up during the mobilisation phase can ask questions on access to the construction site and the duration of project commencement.

Project planning and tracking activities may be reviewed in terms of scope changes, change orders, responsibility allocation, adherence to the schedule, delivery of milestones in the schedule, baseline tracking activities, risk and schedule dependency management. Questions regarding the development process pertain to training and staff development during the project. In complex construction projects, staff may be required to undertake training on health and safety procedures or specialised software such as BIM. The impact of the training and development of staff may provide indications on how future projects may be improved.

The quality and cost management activities must be considered for any setback and applications of quality control and cost overrun causations. The outcome of post-project reviews must lead to further improvement in construction cost estimation, planning, and execution.

9.4 Continuous cost improvement strategies for unique construction projects in Italy

Unique construction projects, as discussed in Chapter 8, consist of large publicly funded projects with invested capital of over $100 million. Bucciol, Chillemi, and Palazzi (2013) studied unique public road projects in Italy and identified the causations of cost overrun as factors related to funding, optimism bias, construction methods, and payments. Plebankiewicz and Wieczorek (2020) also studied a wide range of large projects around the world, including Italy, and echoed the findings of Bucciol et al. (2013). In mitigating the aforementioned cost overrun challenges of unique or large construction projects, the continuous cost improvement strategy of construction material life cycle costing, the circular economy in construction, and constructability assessment form the major improvement stakes. Figure 9.3 illustrates the cost-related issues in large construction projects and how continuous cost-improvement strategies can be used to alleviate the impact of the recognised cost issues on large construction projects. The cost-related issues were extracted from Table 8.4 in Chapter 8.

The idea of continually improving the cost of constructing a large publicly funded project such as the road project in Italy must be based on the strategies of Figures 9.1 and 9.2 and not just the strategies in 9.3. Subsequently, Figure 9.3 presents a trio of continuous improvement strategies that interact to provide a holistic approach to adopting continuous improvement in larger construction projects. The individual components of the strategies in Figure 9.3 are elucidated in Sections 9.4.1–9.4.3.

9.4.1 Circular economy in construction

A circular economy approach in the construction industry is a broad strategy for reducing waste in the construction process and a consistent approach for adopting continuous improvement in construction cost management (Osobajo, Oke, Omotayo, & Obi, 2020). Starting from the design phase, design for deconstruction is an approach to designing facilities to eliminate waste and end-of-life disassembly of structures (Obi, Awuzie, Obi, Omotayo, Oke, & Osobajo, 2021). The circular economy is a system of eliminating waste continually by reusing and repurposing available waste materials. In the construction industry, this process starts with the standardisation of the design process, especially in a factory setting of off-site construction.

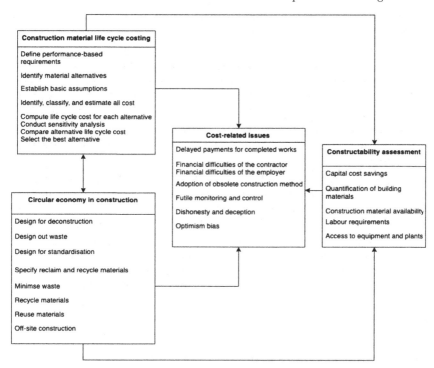

Figure 9.3 Continuous cost improvement strategies in meeting cost-related issues of unique construction projects.

Source: Authors.

The use of pre-cast components eliminates construction and reduces construction cost by up to 70%. The adoption of circular economy practices in construction can eliminate some cost challenges in large public construction projects associated with obsolete construction methods. An example of such obsolete construction method is the use of in-situ concrete when pre-cast components should have been adopted to reduce labour and material cost. The reuse of older materials and waste minimisation are continuous improvement concepts capable of fast-tracking the construction process, thereby mitigating the challenge of financial difficulties on the path of the contractor or employer by reducing construction cost. Circular economy in construction is closely linked with material life cycle costing and constructability assessment. The relationships are explained in the next sections.

9.4.2 *Material life cycle costing*

Material life cycle costing is a subset of life cycle costing. Life cycle costing was discussed in Chapter 3, Section 3.6. In addition to the consideration of operation and maintenance cost, the individual life cycle cost of each construction material adopted in the construction process can be reviewed for its cost-effectiveness. Material life cycle costing aids other continuous cost improvement strategies such as the circular economy and constructability assessment by considering the investment potential of individual construction material (Biolek & Hanák, 2018). Material life cycle costing can be conducted by defining cost problem areas and identifying alternative materials with lower cost and with longer durability. Material life cycle costing is usually carried out in the pre-contract stage of construction.

All cost elements of the construction process can be identified from an elemental cost plan or bills of quantities. In this process, the computation of material life cycle costing must consider the sensitivity analysis for 30 to 60 years. Alternatives are considered after careful comparative analysis. Cost issues in large public projects can benefit from this approach or by selecting the right construction materials, reducing construction cost continually by updating the database, and ensuring timely payments to contractors. Constructability assessment will benefit from the outcome of material life cycle costing when evaluating capital cost savings for large public construction projects.

9.4.3 Constructability assessment

Constructability assessment targets the issue of optimism bias in construction projects. In large public construction projects such as road projects in Italy, the adoption of a constructability assessment can provide an output of potential returns on financial investments (Fadoul, Tizani, & Koch, 2017). The detailed quantification of the building materials cost can be taken from the bills of quantities and material life cycle costing tables.

Construction material availability must be assessed with the option of a circular economy. Labour shortage across Europe is a major problem facing project delivery (Scevik & Vitkova, 2017). Therefore constructability assessment intends to investigate the sources of labour for the construction phase. Alternatively, it can form a major assessment of contractors in the tendering phase of publicly funded projects. Plant and equipment availability should also be assessed as part of the project approval for investment decisions. Constructability assessment, along with the circular economy in the construction industry, can provide up to 14% cost savings in publicly funded projects and significantly lessen optimism bias. In cases where constructability assessment provides negative results to the investors or government, circular economy-related options of off-site or modular construction can form a major part of revaluating the construction material, labour, plant, and equipment. In megaprojects, the aforementioned strategies from 9.2 to 9.4 are very relevant.

9.5 Continuous cost improvement strategies for megaproject in the UK

An example of a megaproject in the UK is the High-Speed Rail 2 (HS2) project between London and the West Midlands. There were cost overruns of over £30 billion in the early stages of the HS2 rail project and some of the factors identified were ground conditions (similar to the context of Figure 9.1) that stemmed from lack of data and models for cost forecasting, technical know-how, poor complexity management, political pressure and economic vagaries such as the exchange rate, inflation, and the importation of materials (Boateng, Chen, & Ogunlana, 2016; Flyvbjerg, 2005; Rothengatter, 2019; Söderlund, Sankaran, & Biesenthal, 2017). The cost issues identified in Figure 9.4 are typical of cost overruns identified in megaprojects and not just HS2 rail projects. In lessening the impact of cost overruns in megaprojects such as the

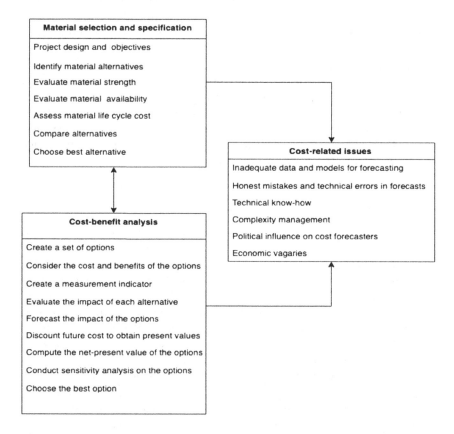

Figure 9.4 Continuous cost improvement strategies in meeting cost-related issues of mega-construction projects.

Source: Authors.

HS2 rail, constructability assessment, circular economy, and material selection and specification are important continuous cost improvement strategies. More specific continuous cost improvement strategies are the use of material selection and specification and the cost-benefit analysis guidelines listed in Figure 9.2.

The components of cost-benefit analysis, materials selection, and specifications strategies are highlighted in Sections 9.5.1 and 9.5.2.

9.5.1 Material selection and specifications

Material selection and a specification strategy for continuous cost improvement in construction are very similar to material life cycle costing, market research, and the use of local materials. The material selection and specification approach is a pre-contract strategy for selecting the best construction material that meets the economic investment and value of a construction project (Chattopadhyay, 2009; Lam, Chan, Poon, Chau, & Chun, 2010). This strategy makes use of the project design and objectives to identify alternative building materials.

Each building material is evaluated for its thermal strength, conductivity, performance, and durability. The purpose of the construction material evaluation is to provide the right specification to meet the needs of cost modelling and forecasting in megaprojects. Consequently, the material life cycle cost is assessed using this approach before choosing the best alternative. This strategy can be combined with a cost-benefit analysis to reduce economic vagaries on megaproject construction.

9.5.2 Cost-benefit analysis

The cost-benefit analysis was discussed in Section 4.4 of Chapter 4 as a construction cost management technique. Cost-benefit analysis is a comprehensive economic appraisal strategy that considers the monetary value of an investment and the financial outcomes of any investment in a project (Robinson, 1996). Megaproject construction must make use of the cost-benefit analysis strategy to evaluate all the material, plant, equipment, construction method, and project options available. Project measurement indicators are then developed to clearly understand the impact of each alternative on the project.

Cost forecasts and models are developed with the aid of discounted cash flow, net present values, and sensitivity analysis. The outcome will lead to the choice of the best construction project alternative. As a continuous improvement strategy, cost-benefit analysis meets the cost issues of political influence, cost modelling, and forecasting, as well as economic vagaries as a pre-contract analysis. The process of cost evaluation can continue throughout the entire project life cycle because a megaproject can be conducted in phases. Hence, each phase can make use of the material selection and specification along with a cost-benefit analysis to determine the cost trajectory of the construction project.

9.6 Summary

The strategies discussed in this chapter are broader strategies for continually mitigating the impact of cost-associated challenges on construction projects. There is an interrelationship between the individual strategies and their applications in alleviating the impact of cost-related issues in the pre-contract and post-contract phases of construction.

Table 9.1 presents a summary of the continuous cost improvement strategies, their stages of application, and selected cost-related issues they are capable of mitigating. The above-mentioned continuous cost improvement strategies have features of a basic continuous improvement approach such as standardisation, waste reduction, heightened communication, modelling, and a cyclical approach to providing enhanced processes in construction. The strategies already exist in the construction industry and this chapter only provided an exposition on how to

Table 9.1 Summary continuous cost improvement strategies

Nr	Strategy	Stage of application	Cost-related issues
1.	Market research	Pre-contract	Project financing and payments
2.	Making use of local materials	Pre-contract	Community engagement issues
3.	Engagement with suppliers and subcontractors	Pre-contract and post-contract	Poor site management
4.	Personal heuristics	Pre-contract and post-contract	Constructability issues
5.	Brainstorming	Pre-contract and post-contract	Design and scope changes
6.	Value engineering	Post-contract	Price escalation
7.	Post-project reviews	Post-contract and end of the project	Errors in construction project cost
8.	Circular economy in construction	Pre-contract and post-contract	Obsolete construction method
9.	Material life cycle costing	Pre-contract and post-contract	Financial difficulties of the contractor and employer
10.	Constructability assessment	Pre-contract and post-contract	Optimism bias
11.	Material selection and specification	Pre-contract	Inadequate data and models for forecasting
12.	Cost-benefit analysis	Pre-contract and post-contract	Economic vagaries

apply each strategy clearly to eliminate or minimise construction cost challenges.

References

Biolek, V. and Hanák, T. (2018). Material life cycle costing of buildings: A case study. *Proceedings of the AIP Conference*, 1978(1), p. 240013. doi: 10.1063/1.5043874

Boateng, P., Chen, Z. and Ogunlana, S. (2016). A dynamic framework for managing the complexities of risks in megaprojects. *International Journal of Technology and Management Research*, 1, pp. 1–13.

Bucciol, A., Chillemi, O. and Palazzi, G. (2013). Cost overrun and auction format in small size public works. *European Journal of Political Economy*, 30, pp. 35–42. doi: 10.1016/j.ejpoleco.2013.01.002

Carrillo, P., Harding, J. and Choudhary, A. (2011). Knowledge discovery from post-project reviews. *Construction Management and Economics*, 29(7), pp. 713–723. doi: 10.1080/01446193.2011.588953

Chattopadhyay, S. (2009). Selection of materials for construction. In *Proceedings of the ASEE Annual Conference and Exposition*. doi: 10.18260/1-2-5119

Cheah, C. Y. J. and Ting, S. K. (2005). Appraisal of value engineering in construction in Southeast Asia. *International Journal of Project Management*, 23(2), pp. 151–158. doi: https://doi.org/10.1016/j.ijproman.2004.07.008

Dada, J. O. and Jagboro, G. O. (2007). An evaluation of the impact of risk on project cost overrun in the Nigerian construction industry. *Journal of Financial Management of Property and Construction*, 12(1), pp. 37–44. doi: 10.1108/13664380780001092

Fadoul, A., Tizani, W. and Koch, C. (2017). Constructability assessment model for buildings design. *Digital Proceedings of the 24th EG-ICE International Workshop on Intelligent Computing in Engineering 2017*, (August), pp. 86–95.

Flyvbjerg, B. (2005). Machiavellian megaprojects. *Antipode*, 37(1), pp. 18–22. doi: 10.1111/j.0066-4812.2005.00471.x

Goodwin, P., Wright, G., & Phillips, L. D. (2004). *Decision analysis for management judgment*. Wiley Chichester.

Ioannidou, D., Meylan, G., Sonnemann, G. and Habert, G. (2017). Is gravel becoming scarce? Evaluating the local criticality of construction aggregates. *Resources, Conservation and Recycling*, 126, pp. 25–33. doi: https://doi.org/10.1016/j.resconrec.2017.07.016

Kern, A. P. and Formoso, C. T. (2006). A model for integrating cost management and production planning and control in construction. *Journal of Financial Management of Property and Construction*, 11(2), pp. 75–90. doi: 10.1108/13664380680001081

Khan, S. A., Kaviani, M.A., Galli, J.B. and Ishtiaq, P. (2019). Application of continuous improvement techniques to improve organization performance. *International Journal of Lean Six Sigma*, 10(2), pp. 542–565. doi: 10.1108/IJLSS-05-2017-0048

Khodeir, L. M. and El Ghandour, A. (2019). Examining the role of value management in controlling cost overrun (application on residential construction projects in Egypt). *Ain Shams Engineering Journal*, 10(3), pp. 471–479. doi: https://doi.org/10.1016/j.asej.2018.11.008

Lam, P. T., Chan, E. H., Poon, C. S., Chau, C. K. and Chun, K. P. (2010). Factors affecting the implementation of green specifications in construction. *Journal of Environmental Management*, 91(3), pp. 654–661. doi: https://doi.org/10.1016/j. jenvman.2009.09.029

Magill, L. J., Jafarifar, N., Watson, A. and Omotayo, T. (2020). 4D BIM integrated construction supply chain logistics to optimise on-site production. *International Journal of Construction Management*, pp. 1–10. doi: 10.1080/ 15623599.2020.1786623

Mancini, L., De Camillis, C. and Pennington, D. (2013). *Security of supply and scarcity of raw materials: Towards a methodological framework for sustainability assessment.* doi: 10.2788/94926

Moyane, S. T. (2018). *Investment decision-making process: heuristics and biases in South African based investment firms* (Doctoral dissertation).

Obi, L., Awuzie, B., Obi, C., Omotayo, T., Oke, A. and Osobajo, O. (2021). BIM for deconstruction: An interpretive structural model of factors influencing implementation. *Buildings*, 11(6), p. 227.

Omotayo, T., Awuzie, B., Egbelakin, T., Obi, L. and Ogunnusi, M. (2020). AHP-systems thinking analyses for Kaizen costing implementation in the construction Industry. *Buildings*, 10(12), 230.

Omotayo, T. and Kulatunga, U. (2017). A Gemba Kaizen model based on BPMN for small- and medium-scale construction businesses in Nigeria. *Journal of Construction Project Management and Innovation*, 7(1), pp. 1760–1778.

Omotayo, T. S., Kulatunga, U. and Bjeirmi, B. (2018). Critical success factors for kaizen implementation in the Nigerian construction industry. *International Journal of Productivity and Performance Management*, 67(9). doi: 10.1108/ IJPPM-11-2017-0296

Osobajo, O. A., Oke A., Omotayo, T. and Obi, L. (2020). A systematic review of circular economy research in the construction industry. *Smart and Sustainable Built Environment*, (ahead-of-print). doi: 10.1108/SASBE-04-2020-0034

Payling, S. J. (1992). Arbitration, perpetual entails and collateral warranties in late-Medieval England: A case study. *The Journal of Legal History*, 13(1), pp. 32–62. doi: 10.1080/01440369208531048

Pennanen, A., Ballard, G. and Haahtela, Y. (2011). Target costing and designing to targets in construction. *Journal of Financial Management of Property and Construction*, 16(1), pp. 52–63. doi: 10.1108/13664381111116089

Plebankiewicz, E. and Wieczorek, D. (2020). Adaptation of a cost overrun risk prediction model to the type of construction facility. *Symmetry*, 12(10), pp. 1–17. doi: 10.3390/sym12101739

Robinson, R. (1996). Economic evaluation and health care decision-making. *Health Policy*, 36(3), pp. 215–229. doi: 10.1016/0168-8510(96)00814-7

Rothengatter, W. (2019). Megaprojects in transportation networks. *Transport Policy*, 75, pp. A1–A15. doi: https://doi.org/10.1016/j.tranpol.2018.08.002

Scevik, V. and Vitkova, E. (2017). Optimization of overhead costs of a construction contract. In Vojko, P. V. & Pavle, K. A. V. (Eds.), *Proceedings of the 26th International Scientific Conference on Economic and Social Development*, Zagreb, pp. 515–523.

Seaker, R. and Waller, M. A. (1996). Brainstorming: The common thread in TQM, empowerment, re-engineering and continuous improvement. *International Journal of Quality and Reliability Management*, 13(1), pp. 24–31. doi: 10.1108/02656719610108305

Söderlund, J., Sankaran, S. and Biesenthal, C. (2017). The past and present of megaprojects. *Project Management Journal*, 48(6), pp. 5–16. doi: 10.1177/875697281704800602

Sprinkle, Z. J. (2018). Heuristics in construction project management. *Angewandte Chemie International Edition*, 6(11), pp. 951–952.

Wijekoon, S. B. and Attanayake, A. M. K. (2012). Study on the cost overruns in road construction projects in Sri Lanka. In *International Conference on Sustainable Built Environment*, Sri Lanka, Paper No 69. The Earls Regency Hotel, Kandy, 14–16 December 2012. Sri Lanka: University of Peradeniya.

Yazdani, A. A. and Tavakkoli-Moghaddam, R. (2012). Integration of the fishbone diagram, brainstorming, and AHP method for problem solving and decision making-a case study. *International Journal of Advanced Manufacturing Technology*, 63(5–8), pp. 651–657. doi: 10.1007/s00170-012-3916-7

SECTION C
CASES

10 Cases on Overhead Cost Reduction and Maintenance through Continuous Improvement

Temitope Omotayo, Udayangani Kulatunga, and Bankole Awuzie

10.1 Introduction

In Chapter 6, overhead costs reduction and maintenance were discussed to present factors leading to higher construction costs and how they can be minimised to leave the contractor with a larger profit margin. Overhead costs and maintenance provisions are essential and cannot be subtracted from the cost build-up of construction projects. Contractors usually include the overhead cost percentage in their "profit and overhead" addition to the final contract sum in the form of an 8%, 10%, or 12% mark-up. Hence, overhead costs are usually approximated mark-up values which will not be known in detail through estimation until the construction process begins. Examples of overhead costs are the cost of mobilisation; setting out; insurance cost; and the cost of electricity, water, and stationery, just to mention a few. In this chapter, the factors influencing construction cost overheads (from Chapter 6) are presented in the form of engaging cases.

Cases on overhead cost reduction through continuous improvement include the following:

- Site conditions
- Construction design economics
- Construction planning and procurement arrangements
- Additional costs
- Economic condition
- Design

Hint for the cases

Continuous improvement of overhead costs involves the continual process of identifying non-value adding activities to reduce and maintain company and project overhead cost but it can never be completely removed from the construction process.
Important reading: Chapters 6 and 9

DOI: 10.1201/9781003176077-13

10.2 Case 1: Site conditions

Site conditions were discussed in Chapter 6, Section 6.3.1, and some underlying drivers were identified. This case looks at a situation where the underground conditions, layout, and shape of the site may impact the overall construction cost.

A purpose-built university student accommodation facility has been proposed for development. The JCT 2016 contract with quantities has been agreed upon by all parties as the standard form of contract for this project. The contract has been awarded and the contractor intends to implement continuous improvement in quantifying the detailed preliminary costs associated with the site conditions. The site survey was produced to indicate the location and shape of the site. The general overhead cost associated with the contract is 9%. A list of preliminary costs has been identified and included in the bills of quantities but these have not been calculated individually. The site survey indicated some underground services and the soil condition. The site mobilisation stage has started and there is a need to create site fencing and offices and to commence excavation. In the process of commencing general excavation of the surfaces, you, the estimator working for the contractor, have been notified of some additional preliminary costs.

a You, the estimator, have identified underground boulders, the removal of which will require additional overhead costs. The estimator has informed the contractor of the need to let the contract administrator who is working for the employer know of the situations to claim for relevant events from the contract. Whatever the case, the contractor will still incur some overhead cost related to plant hire, delays, and unproductive hours. You are aware of the continuous improvement process and what is expected of you. What will you do in this situation? How will you reduce the impact of the delays on the construction cost? Which of the continuous improvement steps will you adopt to mitigate the impact of unidentified rock boulders on the site on the overhead cost? Considering the existence of underground boulders, how will you maintain the overhead costs associated with the extra activities on-site?

b The setting out or laying out process is an important process in transferring the information from the drawing to the site. This process will make use of lines and peg, drawings, and more importantly the experience of the on-site labourers and construction professionals such as the contractor, architect, and construction project manager. The process of setting out involves general excavation, topsoil removal, and trench excavations. Column pit excavations will be excavated for larger buildings requiring structural column bases. Within the project described in this case, a mistake has been made in the process of transferring the information from the drawings to the construction site. One of the ground floor rooms has not been excavated for and the ground condition on this site will require rock removal. This mistake will require additional cost, time, labour, and plant resources. What will you do in this situation? Will you inform the employer and contract administrator? Who would you consider is to blame in this situation? How would you quantify the additional cost of construction? What continuous improvement approaches would you adopt to manage the key construction stakeholders on this project? Can you classify the types of continuous cost improvement mechanisms that can reduce or mitigate the impact of this mistake on the overall construction process?

---

c The undulating shape of the site is making it difficult for the construction of external works involving a car park, landscaping, and underground services such as the manholes, drains, and gas pipes. A portion of the land is higher and, in some areas, it is lower. Bulk excavations and fillings have been recommended by the construction project manager. Although the survey plan indicated the undulating topography, additional costs will be required to deliver the external works. Are there any continuous cost improvement approaches that may be adopted to reduce and maintain the impact of the bulk excavations and fillings on the overall construction cost? Do you think there is a need for detailed measurement of additional earthworks? What will be the impact of measurement of drainage and services below ground on the overall construction cost? How will you address the challenge of keeping the overhead cost mark-up within the limits of 9%?

10.3 Case 2: Construction design economics

Construction design economics is a precontract factor capable of determining the overall direction of a construction project. In Chapter 6, Section 6.3.2, design economics was highlighted. The underlying drivers are the gross floor area, the project construction estimate, the nature of the project, the schedule of the project, and the proposed building height and shape of the structure to be constructed. These aforementioned drivers have been considered in the development of a car park. The overhead cost considerations in the design economics of the car park adopted the designing-to-cost approach to provide a tailor-fitted five-storey car park of 3,433m². The following problems were detected in the design and cost planning stages of the project:

a The gross internal floor area (GIFA) of the car park is 3,433m². However, the cost per m² produced a total cost higher than the budget the client had earlier proposed and agreed on for the design because of economic vagaries such as inflation, the exchange rate, and the cost of importing the building materials. In light of these challenges, the GIFA must be reduced to meet the financial needs of the employer. This is still possible because you are still in the design and cost planning stage. What are the steps to take to reduce the GIFA? What will be the likely GIFA if the existing cost per m² is £2,584/m²? How will you ensure that this does not affect the morale of the architect and cost planner involved in the revision? Lastly, will you consider this to be a continuous cost improvement process?

---

b You are working for a contractor who intends to bid for the car park construction. The project construction estimated budget has been produced in the form of a bill of quantities for competitive tender purposes. However, upon careful revision of the bill of quantities, you have discovered a computational error in the final value of construction and the percentage of the overhead cost will not cover the company and project overhead costs. The question of how to adopt the right mark-up fully for the overhead cost is a major concern because of the uncertainties in the economy. The estimated budget has also included a contingency sum which your organisation does not consider as an important part of the project until it is required. You have corrected the computational errors discovered in the bill of quantities, but what steps will you take to compute the right overhead cost mark-up for this project? Do you think it is essential for you to identify how to reduce non-value-adding activities in the project and design an approach to limit physical and non-physical waste in the construction process?

---

c The schedule of the car park project in the form of a Gantt chart with activity network diagrams has been produced as part of the tendering process. However, considering the existing construction cost, you find it difficult to reconcile the Gantt chart with the bills of quantities because the employer wants this project delivered within 24 months. The estimated duration for the project is 27 months but the contract has agreed to deliver the project within 24 months. You have been asked to devise a means of reallocation of the resources to fit the requested schedule. How will the overhead costs be impacted in the face of this request? Do you think there will be higher overhead costs than originally planned? How can the implementation of continuous cost improvement measures reduce the project schedule by three months?

10.4 Case 3: Construction planning and procurement arrangements

Construction planning and procurement arrangement have a way of influencing the nature and percentage allocated to overheads. Continuous improvement of the overhead costs was discussed in Section 6.3.3 of Chapter 6. A case of a refinery construction with multiple procurements approaches such as construction management, design and build, and traditional procurement strategies created some level of complexity concerning how overhead costs will be experienced in this megaproject. The following scenarios were identified for your contemplation and knowledge:

a The road development around the refinery construction, the 3D design, and bills of quantities were fed into a common data environment where all the stakeholders such as the employer, civil engineer, architect, contract, construction project manager, and subcontractors can view it. This was made possible because of the use of BIM. However, an architect who was not part of the design process of the road design has detected a clash between the external works of one of the buildings and the road network. This will affect the 5D BIM information and other overhead costs which are not linked directly with the road project. Some of the construction activities have started. How will you address the impact of the changes on the overhead costs as the cost planner? The information fed to you earlier was incomplete and did not provide the final level of design for the road before it was used by other stakeholders involved in the project. How can this be resolved in the future through a continuous improvement strategy?

----

b Construction methods

The off-site construction method was applied as the sole construction method, and pre-cast concrete components and refinery modules were transported from various sites in and outside the country. The cost of transportation has increased after the exchange rate fluctuated because of the recent elections. This affected the import duty on the refinery module. Similarly, the cost of petrol has risen more than expected and the transportation overhead cost also seems to be higher than expected. This higher overhead cost is very significant and must be managed to keep the construction project within the budgeted benchmark. Which continuous cost improvement strategy can you implement to mitigate the impact of higher transportation cost on this megaproject? Who are the construction stakeholders to invite for an important meeting to address this urgent rise in overhead cost? Identify a key continuous cost improvement strategy to keep the overhead cost within the desired limit.

_____

_____

_____

_____

c The quality requirements for the refinery modules and building projects were produced as part of the tender documents. This was produced in the form of specifications and materials selected from a database. As you were involved in the construction process, you discovered that the higher quality of cladding materials for the building will require the work of a specialist outside the scope of your construction company. Hence, the specialist will have to provide, supply, and install the external cladding you have selected. This will lead to an extra overhead cost from administrative charges and payments. Is there a possibility of eliminating this overhead cost without changing the quality requirement? If you go down this path of eliminating the use of a specialist contractor, how will this affect the overall construction cost and the performance of the project?

_____

_____

_____

d The community where the proposed refinery will be situated has staged a protest. The protest is about the environmental impact the refinery will have on the residents, their non-inclusion in the construction process, and the negative impact of the construction on traffic congestion. Further protests will lead to delay in the construction project and additional overhead costs. You have been asked to review the procurement strategy adopted to engage the community in a discussion to further understand their requests. You have considered their request and realised that some continuous cost improvement strategies will be important in the overall procurement process. Itemise and discuss some continuous improvement strategies which may be essential in engaging the local community in the construction process, further improving the construction cost.

---

e The contractors have not been paid by the employer and many subcontractors are also experiencing a delay in payment. The clauses relevant to the standard form of contract have been evoked by the contractors to suspend construction activities on the site. Do you think the contractor has made the right decision? In the event of this delayed payment, the employer has contracted you to suggest alternative construction stakeholder management approaches to foster effective communication with the contractors. Is there a continuous improvement strategy that can be used to resolve this challenge?

10.5 Case 4: Additional costs

Additional costs of construction are very common in construction projects. When there are additional costs, there will be additional construction overheads. Examples of additional costs as indicated in Chapter 6, Section 6.3.4 are variations and rework, plant and equipment maintenance, fuel price, environmental cost, plan depreciation cost, and material handling and delivery. A hospital construction project involving many buildings and multiple contractors using the management contracting route has experienced some additional costs, which are explained in the sections below:

a Variations were discovered after discrepancies in the external wall construction were misinterpreted from the drawing. This is a major variation that will require rework from the contractor. As part of the cost planning team working with the contractor, you have been asked to evaluate the impact of the variation on the contractor's profit margin and the project overhead costs. Continuous cost improvement strategies are available for you to reduce the impact of this variation on the overall performance of the construction project, to manage essential stakeholders, and to propose a new measure of preventing a reoccurrence. Which continuous cost improvement strategy will be useful to fulfil this request? How can continuous cost improvement strategies ensure a consistent approach to reducing construction variations? The management contracting procurement route makes use of multiple work package contractors to deliver various aspects and elements of the building concurrently. What will be the impact of the variation order on the delivery of other aspects of the hospital?

b   The plant and equipment such as tree pullers, 360-degree excavators, the concrete mixer, loading shovel, forward tipping dumper, and lorry loader are being used on the site. Some of the equipment was hired by the contractors. In the event of a major variation on the construction site, some of the concurrent activities were delayed and this led to higher overhead costs and potential depreciation costs. Some 360-degree excavators require servicing and another similar plant will not be used while the variation is still under assessment. Which continuous cost improvement strategies can lighten the impact of the additional cost of plant hire, downtime, and maintenance? How can you continually limit the influence of non-productive hours on the overhead costs?

c   Double glazed windows were supplied by a subcontractor and you have decided to make use of your ad hoc labour to install them. In the process of handling the double-glazed windows, five of them were damaged and cracks were discovered in the glazing. This is an additional cost for which the contractor is liable. Now, you have thought about the process of delivery and supply, which should have been supported by fixing. Thus, there should have been a supply and fix approach for installing the windows in the building. Asides from the supply and fix approach, is there a continuous cost improvement approach to reduce waste in this scenario? How would you embrace a more careful method towards reducing wastage in the supply and installation of double-glazed windows with continuous improvement?

d   The high price of diesel fuel in the country has added to the overhead cost of maintaining the hired plant on site. There is no other option than to buy the diesel at the market price and this will add to the existing overhead cost. In planning for the future, are there other alternative sources of fuel or plant which are sustainable for the construction process? How will you ensure that the additional cost of fuel does not have much impact on the overhead cost of construction?

_____

_____

_____

_____

10.6 Case 5: Economic condition

Overhead costs must be managed effectively with continuous improvement. However, there are some unpredictable exogenous influences on construction processes and cost, which may be very difficult to manage. Section 6.3.5 in Chapter 6 summarised some of the economic influences on overhead cost as inflation, economic forces, the economy of the site, construction output, interest and exchange rates. In the development of a new 0.5-kilometre bridge over a period of 16 months, the construction materials were imported and some economic conditions influenced the delivery of the construction materials and the construction process.

a The bill of quantities and estimates prepared for the construction of the bridge include an inflation rate of 3% when the project is expected to start in the third quarter of the year. In the fourth quarter of the year, the inflation rate rose to 4% and the dynamic effect of this change impacted the price of cement, pre-cast concrete components, and temporary formwork. Considering the overall cost of construction, the supplier and subcontractor's cost and related overheads will increase. What type of continuous cost improvement strategy can potentially maintain the overhead cost? How will you maintain the percentage allocated to overheads so that it will not influence the construction cost negatively?

_____

_____

_____

_____

b The importation of steel from China is being affected by an unfavourable currency exchange rate which will have a ripple effect on the construction cost and processes. From a broader perspective, the changes in the exchange rate may lead to a cost overrun of about 10% if alternatives are not used. Continuous cost improvement may be difficult to adopt in this scenario. However, there are similar strategies on the construction methods and technologies which may assuage the negative influence of an unfavourable exchange rate. What are the plausible continuous cost improvement strategies, technologies, or construction methods that can mitigate the impact of the exchange rate on the bridge construction project? Similarly, in what ways will an unfavourable exchange rate affect other facets of the bridge construction project?

_____

_____

_____

_____

c The location of the bridge construction site experienced traffic diversion in a busy township. The traffic diversions and the cost of engaging in a construction project in a busy environment constitute more overhead costs. The contactor has detected additional overhead costs resulting from temporary cofferdams over the water to create the piers. Which of the continuous cost improvement strategies would you use in the event of emerging overhead costs which were not included in the contract sum? Do you think you can learn lessons from previous bridge construction projects in a busy town centre?

_____

_____

_____

_____

10.7 Case 6: Design

The design of any structure affects the cost variables incorporating a construction project. More importantly, constructability, errors, and deficiencies in designs are some of the limitations, leading to higher overhead cost and final construction cost. Section 6.3.6 of Chapter 6 explained how design factors drive the direction of project costs. In the proposed construction of a shopping mall in a busy city centre, the challenges of constructability and design implementation were encountered as explained in the sections below:

a The design of the shopping mall, which will feature large departmental stores, had building plans with circular and pyramid shapes. The specialists are expected to deliver the project under the direct design and build procurement route. The procurement option adopted in this project requires multiple subcontractors to provide a fast-track delivery of the shopping mall ahead of the Christmas holidays. The design produced for this project has some complexities that will specifically require the services of specialist contractors to deliver. How would you assess the constructability of the design? What processes and approaches will you adopt to diminish cost overruns effectively? Are there particular overhead costs in this project to identify?

_____

_____

_____

_____

b Using the abovementioned project, briefly explain how you will manage the stakeholders to rectify one of the incomplete designs. You have detected an incomplete section plan as part of the drawings provided. Are there new design and construction options that can be used to resolve incomplete sections in designs easily? What steps will you take to resolve the existence of an incomplete design in a construction project? Do you think there are additional overhead costs involved in resolving incomplete designs? If you have identified overhead costs, how do you intend to use continuous improvement to reduce the influence of this additional overhead cost on the construction process? Finally, what are the implications of an incomplete design on the construction process?

c Just as there were inadequacies in the design used in the shopping mall construction, some errors in the structural detail for the columns and beams were identified by the contractor. These errors pertain to the size of the columns and beams. The cost planning team did not identify these errors and made use of the wrong sizes to produce the bills of quantities. These quantities were transferred to the construction project and the construction project manager identified the errors during the construction phase. A variation order was issued and the resultant effect of this will lead to additional overhead costs, plant downtime, and a meeting with the contract administrator. What are the additional overhead costs derived from the errors in this design? What processes or strategies can contribute to effective overhead cost management?

10.8 Summary

The overhead cost factors presented in Chapter 6 are discussed with relevant cases in this chapter. Continuous improvement of the overhead cost of construction can be achieved with the strategies mentioned in Chapter 9. Modern construction activities must be based on modern methods of construction and practices because the dynamics of challenges influencing construction works are increasingly becoming more demanding.

11 Cases on Continuous Cost Improvement Attributes for Monitoring Construction Projects – PART I

Temitope Omotayo, Udayangani Kulatunga, and Bankole Awuzie

11.1 Introduction

There are several essential construction cost management techniques that are used to monitor and control construction costs. Some of these techniques have been discussed in Chapters 1, 4, 5, and 6. Variation management, monitoring of costs of plant and construction materials, supplier cost, and taking corrective action are some of the construction cost controlling techniques. The use of site meetings and risk management in the construction process can also drive down costs. Continuous improvement practices in cost controlling have been documented as being important in the construction process. In this chapter, cases on continuous cost improvement using cost controlling techniques are presented with applications within specific projects.

Construction cost monitoring and controlling techniques can be merged with the plan-do-check-act (PDCA) technique for continuous improvement by identifying gaps in extant strategies for reducing production cost. Hence, continuous cost improvement in the construction industry adapts the changes in the construction process to identify opportunities for reducing construction cost using various cost monitoring and controlling techniques. An elemental cost plan/breakdown of construction must be produced to inform the calculation process. The formulas for reducing construction cost make use of the total cost of production divided by the actual cost. The next step should be to calculate the total actual cost of constructing each element in the project by multiplying the actual elemental cost of construction by the estimated production (quantities of elements produced) for each section of the elements.

1 Per actual product cost from the construction process
$$= \frac{\text{Total actual cost of element}}{\text{The actual production of element}}$$

2 The estimated amount of total actual cost = Per actual element cost of construction [1] × estimated production for the element

DOI: 10.1201/9781003176077-14

3 Continuous improvement cost target for construction project = Estimated amount of total construction actual cost [2] × ratio of the cost reduction target

4 Assignment elemental cost

$$= \frac{\text{Cost directly controlled in single plant/material/labour}}{\text{Cost directly controlled in all plants/material/labour}}$$

5 Continuous improvement cost target for each element = Continuous improvement cost target for the construction project [3] × assignment ratio [4]

Adapted from Monden and Hamada (1991).

The third step that intends to calculate the continuous improvement cost target for the construction project will involve a multiplication of the estimated total construction cost by the ratio of cost reduction target. The ratio of cost reduction targets can be derived from Figure 1.2 in Chapter 1. The actual elemental cost of construction is calculated by dividing the single costs directly controlled in terms of the plant, labour, and materials against all the costs directly controlled when using the plant, labour, and materials. The final step will break down each element's continuous improvement cost target by multiplying the continuous improvement cost target for the construction project by the assignment ratio. Other techniques involving the PDCA continuous cost improvement approach will be elucidated with case examples of how construction projects can practically make continuous improvement in construction processes. Some of these cost monitoring and controlling techniques in the construction industry have been adopted without mentioning "continuous cost improvement". Therefore, some of the techniques discussed in this chapter are typical post-contract cost controlling techniques presented in terms of applicability in continuous cost improvement. Continuous cost improvement monitoring and controlling techniques have been borrowed from the Toyota Production System (TPS) and manufacturing processes in Japan.

The individual cases on continuous cost improvement attributes for monitoring construction projects will focus on the following:

- Variation management
- Plant hire cost
- Labour cost
- Administrative charges
- Monitoring construction material cost
- All forms of risk
- Financial risk
- Drawing reviews
- Meeting the benchmark cost for each element
- Taking immediate corrective action

- Managing subcontractors and suppliers' cost
- The outcome of weekly/monthly site meetings

Hint for the cases

It is easier to monitor and manage construction processes than to control them. Continuous cost improvement practices can use cost targets to reduce incidents of unnecessary expenditure, attain productivity efficiency, reduce risks and cost overruns.
Important reading: Chapters 1, 4, 5, and 6

11.2 Case 1: Variation management

Variation management is a construction cost monitoring technique capable of deciding the course of a construction project. Construction variations may be minor or major. Examples of variations and variation management have been provided in Chapter 4, Section 4.7.

The process of reducing construction cost overrun must be predicated on cost monitoring and variation management. You are the new construction manager in charge of the construction of a cinema and restaurant construction project. The project budget is £1.3 million, and the organisation you work with uses a construction management strategy to procure this new structure. The JCT 2016 with quantities has been selected as the standard form of contract for this construction project. You have supervised the construction of the substructure, which was delivered by a subcontractor who specialises in pad foundations and structural steelworks. After the first interim valuation, you discovered the following variations in the actual work completed on site:

a You have realised that the subcontractor has used the lowest grade of in-situ concrete, and this may affect the structural integrity of the building. You have raised this with the subcontractor as a major variation, and rework has been issued. The subcontractor blamed the supplier for the lower quality of concrete delivered on the site. You have revalued the change order in this variation and understood the financial constraints the subcontractor will experience in the rework. Based on your experience with the application of continuous cost improvement, how are you going to apply the PDCA principle in this scenario? Are there new approaches by which you may advise the subcontractor to reduce the construction cost while maintaining the quality for an expected profit margin?

---

b In the same construction project, you are responsible for the construction and cost monitoring of the project. The rework resulting from the variation is important and must be completed before the construction progresses. The rework will require an extension of time and therefore lead to additional costs related to construction activities. The extension of time will incur an additional 5% of the existing construction cost. Do you think you can compute a continuous cost improvement target for the fundamental elements? What cost gaps can you reduce in the process of the rework?

11.3 Case 2: Plant hire cost

Your company has been awarded a contract, and you are the construction manager involved in constructing an office complex in a city centre. The quantity surveyor has produced the elemental cost plan and the bill of quantities. The total construction cost is £3.2 million, and the quantity surveyors have factored in the cost of plant hire in a detailed estimate.

a You intend to hire an excavator, tower crane, and another related plant for the construction activities you will not be subcontracting. The entire plant hiring cost accounts for 12% of the construction cost, and you are seeking opportunities to reduce the cost by 3% to add to the estimated profit margin of 10% in order to provide an estimated profit margin of 13%. What are the cost variables associated with plant hire cost that must be identified to reduce the cost of hiring and maintaining excavators, a tower crane, and another plant you intend to hire? After identifying the cost factor for the plant hire, can you investigate alternatives to the existing plant and other cheaper plant sources for the construction project?

----

b After hiring the plant, you require for the construction project, and you realised that you had hired the wrong size of concrete plant, which will be too large for the construction project. The larger concrete plant will imply additional construction cost. You have decided to return this large concrete plant after two months of use. You realised that this was a mistake that led to the unnecessary cost of plant hire and maintenance. The rental rates are very high, and you may have to consider lower rental rates to meet the profit target for your company.

What should have been done before hiring the plant? Is there another plant that you should have hired that may lead to cost leakages? Can you adopt the PDCA process in plant hire cost reduction? If the answer is "Yes", what are the steps you must take to apply PDCA in plant hire cost reduction?

_____

_____

_____

_____

11.4 Case 3: Labour cost

Labour cost was discussed in Chapter 4, Section 4.3. In terms of continuous cost improvement, labour wages must not be reduced. However, labour cost may be reduced by optimising the construction schedules and incentivising high productivity and the use of offsite construction. In labour cost reduction, ethical construction standards of not short-changing on-site labour must be adhered to. By reducing the labour hours on-site through one of the modern construction methods discussed in Chapter 5, labour costs can be reduced. Hence, continuous improvement in labour cost can be attained through modern construction methods like 3D-printed houses, cross-laminated timber, predicting the behaviour and productivity of labour on-site and off-site, and modular construction.

a A contract intends to submit a proposal for a new primary school's design and build construction project. The contract intends to cut labour cost by 60% and make use of an innovative construction method, thereby providing an opportunity for the contractor to gain up to 18% to 20% profit in the project. First, you to assess the feasibility of reducing labour cost by 60% and providing a profit margin of up to 20% with a modern construction method. Which method will you consider? How will you provide the labour cost reduction strategy that can meet the needs of the contractor? What will be the impact of reducing labour cost on the construction industry development with a modern construction method?

_____

_____

_____

_____

b In a similar construction project, the design and build approach has been adopted. This project made use of structural steelworks comprising steel columns and beams. The contractor labour cost has been estimated, and there is a labour shortage to deliver the construction project. The project manager will need an additional ten skilled labourers to work on the steel structures and does not intend to subcontract this phase of work because the subcontractors also have a shortage of labour. The shortage of labour implies that the existing skilled labourers on site will work longer hours and therefore be paid more than expected. In this situation, you are free to explore other ways of reducing construction labour hours while maintaining wages. There are options for you to adjust the nature of construction components used for the external walls and frame. You are also expected to identify and propose changes to the labour schedule. Identify continuous cost improvement methods for reducing the labour hours.

11.5 Case 4: Administrative charges

Chapter 6 discussed overhead costs emanating from preliminary works such as electricity, water, supply, insurance, and many other unmeasurable construction components. Likewise, office overhead costs were associated with company overhead costs. Administrative charges remain one of the significant overhead costs. These can be derived from overhead costs associated with the payment of labourers, subcontractors, and suppliers. Administrative charges may be charged by the insurers for payments to subcontractors, depreciation cost, suppliers, electricity usage, or wages. In some instances, administrative charges may be up to 15%.

A contractor is responsible for constructing a multi-storey block of flats. However, the site was not handed over to this contractor on time. There is a three-week delay in taking possession of the construction site, and the contractor is claiming additional cost for administrative changes related to wage payment, plant hire, electricity usage, the insurance company, and another associated overhead cost. The contactor is claiming an additional 15% for administrative charges. The client has declined to make this payment after the contractor has finally taken possession of the site. This is a dispute which should have been avoided if the client had handed over the construction site on time. Are there continuous improvement practices the client could have taken to ensure that the contractor took possession of the site? Equally, are there alternative means of reducing administrative changes using modern methods of construction?

11.6 Case 5: Monitoring construction material cost

The cost of construction materials contributes up to 68% to 70% of the construction cost. Chapters 1, 4, 5, and 6 discussed the elements of construction cost, modern methods of construction, and how continuous improvement can be applied to either reduce or maintain construction cost. Innovative ways of reducing construction cost through modern construction methods can receive a boost with the application of continuous improvement. By monitoring construction cost, material waste, and handling and delivery, continuous improvement can be implemented in the construction process through PDCA.

a   In an IT construction project management activity involving the smart procurement classroom in a university, a specialist contractor was awarded the contract to supply and fix smart boards and other IT equipment such as the projector, networking system, and relevant IT infrastructure. The cost of the project was £563,000. The contractor made use of previous designs used on a similar project completed a year earlier. In this situation where a previous design was used, the lessons learned from the project were not reviewed, and the specialist contractor had more challenges on this new smart building project because the prices of smart equipment such as the smartboard had increased by 8%. The specialist contractor was faced with the challenge of cost overrun. How should this specialist contractor have improved the delivery of construction cost for this smart classroom building?

b   In the same construction of a smart classroom, the subcontractor in charge of the external wall construction has just received information of an increment in the cost of bricks and cement. This subcontractor has notified the main contractor of the changes in the cost of building materials. How will this affect their overheads and profits? There are several alternatives to bricks and cement that can be explored to deliver the project; however, it is necessary to evaluate the cost targets of the alternatives. The subcontractor has contacted you to provide a continuous cost target for the new choice of external walls other than bricks. Therefore, provide an example of alternatives to bricks and cement along with their continuous cost improvement targets.

11.7 Case 6: All forms of risks

Risk management cannot be ignored in any construction project. The risk identification, evaluation, assessment, mitigation, control, and ownership can be aligned with the ideals of continuous improvement to reduce incidents of cost overrun. Section 4.6 of Chapter 4 addressed the topic of identifying cost overruns in construction projects. Risk management in megaproject delivery, just as every other scale of a construction project, is a continuous process that must be integrated with cost management.

Failure to continue the risk management process throughout the project lifecycle can lead to the emergence of risks identified in the register. An example of this was the case of the rail project with a budget of £1.6 billion. The civil engineer and construction project manager had prepared a risk register that associated ownership with specific stakeholders in the project, including the contractor. The cost was allocated to each identified risk. It was the contractor's duty to monitor the impact of the health and safety involved in hoisting and placing the rails on the construction project. In the execution phase of this project, one of the construction workers (working for the contractor) sustained a serious injury which could have been avoided if the contractor had taken the precaution to update the risk register and inform the workers on-site when there was an equipment change. What are the implications of poor continuous improvement in risks management on this project? What were the reasonable steps the contractor should have taken in this project? How can continuous improvement and risk management work together to reduce construction cost?

11.8 Case 7: Financial risk

Financial risk management in construction considers how different scenarios may cause overruns in a construction project. In the same project as Section 11.7, the cost implications were also attached to the risk management plan when the risks were identified. In smaller projects, a contingency fund may be divided across the identified risks. However, financial risk assessment is not only provided for a construction project but for the contractor and the employer. The construction of the rail lines in Section 11.7 made use of a contractor who went into liquidation in the middle of the project schedule. The contractor was selected through the competitive tendering for design and building, and all necessary references and checklist requirements were met. However, it was not envisaged that the contractor would go out of business before the end of the project. Are there any continuous cost improvement strategies essential for contractor selection? What advice will you give to the employer to resolve this challenge?

11.9 Case 8: Drawing reviews

Drawing reviews are important in constructability assessment before the award of a contract. Modern methods such as BIM and other methods, as discussed in Chapter 5, provide multiple opportunities to reduce errors in construction, alleviate cost overruns and provide an opportunity for continuous cost improvement. Drawing reviews must be conducted effectively before construction cost improvement. Mistakes in drawings will provide the wrong continuous cost target for construction elements.

a   An architect in a BIM-driven project has produced 3D designs for a 5D cost planning process. The architect claims to have provided a design with the highest level of development. The quantity surveyor in charge of the 5D cost planning process has detected a clash between the structural steel beams and the plumbing services in the design. The clash will require major adjustments to the design and quantities of the pipe works and steel beams. This was not detected earlier in the 4D scheduling. How are the implications of the clash detected in the design? How can you relate this process with the continuous cost improvement agenda of eliminating waste and creating cost targets?

b   The quantity surveyor detected other mistakes in the electrical design of the 3D section plans. The electrical engineer produced the designs after the building elevation, plans, and sections had been produced by the architect. A clash was detected in the positioning of the electrical cables along the lines of the plumbing. Similarly, containments were not used to conceal the cables where necessary. The changes in drawings reviews will delay the final tender documentation, and the employer must be informed of the delays. What are the key implications of this drawing review on the progress of the construction project? Is this a continuous improvement task? How will this influence the cost from the perspective of the employer?

11.10 Case 9: Meeting the cost benchmark of individual construction elements

The core element of continuous cost improvement is benchmarking the cost of individual construction elements. Cost targets are produced as the basis of continuous improvement. However, several factors such as fluctuations, inflation, financial risks, shortage of labour, and professional expertise keep to the cost benchmark.

a A benchmark was set for the cost of plant hire, depreciation, and maintenance. Similar benchmarks were set for each cost element in the new housing development. The overhead costs experienced price fluctuations and inflation of 2.1% in the 5th of the 18 months of the duration of the construction project. The housing development consists of 22 houses, and the increment in inflation will result in the cost spiralling beyond the overall benchmark. What approaches must the construction project management and cost planning team apply to reduce price fluctuations and inflation on the construction project?

_____

_____

_____

_____

b In a similar scenario as above, where continuous cost improvement has been applied in the construction project, and the economic vagaries are proving to weaken this lean thinking strategy, are there alternatives to applying continuous improvement effectively in the construction project? It is evident that continuous cost improvement is not perfect even with benchmarking, and there is no single-point solution to existing problems facing construction project costs. Recommend a combination of construction cost management systems combined with continuous cost improvement to minimise the impact of economic challenges on cost benchmarking.

_____

_____

_____

_____

11.11 Case 10: Taking immediate corrective action

Taking corrective action is one of the ways of preventing cost overruns in construction projects. It is also a continuous improvement approach to proactively ensuring that waste does not occur and when it does, rectifications are immediately applied. Taking corrective action was discussed in Chapter 4, Sections 4.6 and 4.10 under the headings of identifying cost overruns and the PDCA principle.

a A social housing remedial effort to refurbish multiple houses has been set up with the aid of a building surveyor, quantity surveyor, and facilities manager. The building surveyor has identified and documented a schedule of building elements requiring refurbishment. The quantity surveyor has provided associated costs on the project. The building surveyor was informed that further cracks in the bathroom walls and soil vent pipes that must be replaced had not been identified. The cost in the dilapidation schedule did not identify all defects. The overall budget of the project has been drawn up and an additional £5,000 must be added to the budgeted cost. What steps should the building surveyor have taken to ensure this is resolved using continuous improvement?

_____

_____

_____

_____

b In another new build housing development, the joiner has made a mistake in providing a kitchen cabinet that is obstructing the door's pathway. The contactor detected this mistake during an inspection exercise. This mistake seemed trivial and was not corrected because the tilers were working to meet their deadlines. What will be the implications of the triviality of this mistake on the overall construction process? Do you think the contractor has made a mistake in overlooking this mistake which further delayed the construction project? What are the continuous cost improvement steps to take to correct this mistake?

_____

_____

_____

_____

11.12 Case 11: Managing subcontractor and suppliers' cost

Subcontractor and suppliers' cost can be managed effectively to foster trust in the construction process. In instances where subcontractors and suppliers are not paid on time, there may be distrust and further delays in the delivery of construction materials or aspects of the project required of the subcontractor. Correspondingly, subcontractor and suppliers' cost are part of the overheads incurred by the main contractor. They include administrative charges associated with payments, value-added tax (VAT), and bank charges.

a A general contractor negotiated to make use of the fixed price-lump sum contract with the employer. At the construction of the partition walls, the subcontractor responsible for the electrical installation in the building has complained of late payment. The contractor has also informed the subcontractor of late payments from the employer. This leads to a distrust issue as there is no third-party contract between the subcontractor and the employer. Before this period, the contractor had complained of increasing overheads in the payment mechanism adopted. How should this contractor manage the subcontractor and rebuild the declining levels of trust existing between them? What continuous cost improvement methods are required to eliminate or reduce the subcontractor's costs?

---

b   During this same project, a supplier imported and delivered gravel to the construction site for the external works. After the gravel delivery, the payment was made. Two days later, the contractor realised that if they had chosen a local supplier, they would have reduced the supplier's cost by almost 25%. Which continuous cost improvement strategy could have been adopted in this regard?

11.13 Case 12: Outcome of weekly/monthly site meetings

In Chapter 4, Section 4.8 the technique of site meeting in construction cost management was discussed. Weekly or monthly meetings on a construction project are a major tool for continuous improvement. The construction project manager needs to adopt continuous improvement in the cost management process and further support the philosophy backing the concept by improving communication during construction. Similarly, weekly or monthly site meetings are opportunities for continuous improvement in the construction project and not just cost management. A site meeting was held to review the interim valuations of a construction project. The contractor has reviewed the cost value reconciliation register, and the recent projections indicate that the construction project will lead to financial losses. How should the contractor make use of the weekly site meeting to reduce the projected losses? Which continuous improvement strategy can be employed to mitigate the impact of cost overrun through the site meetings?

11.14 Summary

Continuous cost improvement techniques must be applied along with the cost target benchmarking for each element and strategy. For instance, a site meeting is required to review and identify problem areas in a construction process. Site meetings also provide a broader approach to evaluating the cost targets using the PDCA principle. Site meetings can lead to taking corrective action and effectively managing subcontractor and supplier's costs. It is also essential to monitor the cost benchmarks of individual elements. Drawing reviews before the cost planning process are important and should be a continuous improvement process with the application of modern construction methods such as BIM. The application of continuous financial risk management throughout the project must be monitoring labour, material, plant, and administrative charges. Variation management must address any rework or variation to keep the individual cost elements within established continuous cost benchmarks. Overall, continuous improvement must continuously identify alternative means of reducing construction cost throughout the lifecycle of construction projects.

Reference

Monden, Y. and Hamada, K. (1991). Target costing and kaizen costing in Japanese automobile companies. *Journal of Management Accounting Research*, 3, pp. 16–34.

12 Cases on Continuous Cost Improvement Attributes in the Construction Industry – PART II

Temitope Omotayo, Udayangani Kulatunga, and Bankole Awuzie

12.1 Introduction

This chapter highlights more cases on project practices in furtherance of presenting continuous cost improvement attributes in the construction industry. Chapters 1, 4, 5, 6, and 9 have provided a background of techniques and strategies for adopting continuous improvement in the construction industry. In Chapter 11, variation management, plant hire cost, labour cost, and weekly or monthly site meetings were highlighted in some of the cases presented. This chapter provides more examples of price fluctuations, changes in laws, constructability, norm checks, wastage checks, alternative technology, facilities management, power and water consumption, long-term maintenance, and production continuity. The core principle of PDCA interplays with the activities mentioned above on how continuous cost improvement is practised in the construction industry. Some of the cases below overlap with each other. However, the examples emphasise the need to improve construction outputs and cost continuously.

Cases on construction activities and attributes that must be managed for continuous improvement:

- Price fluctuation
- Changes in laws and regulations
- Compliance with laws
- Constructability or buildability check
- Design economics norms check
- Cost norms check
- Wastage check
- Alternative technology check
- Facility management issues
- Long-term maintenance agreement drafts
- Electric power and water consumption check
- Production continuity

DOI: 10.1201/9781003176077-15

Hint for the cases

New trends in government regulations, laws, technology, and construction must prompt construction professionals to react to improvements proactively.
  Important reading: Chapters 1, 4, 5, 6, and 9

12.2 Case 13: Price fluctuation

Price fluctuations of construction materials such as cement, concrete, steel rods, timber, and gravel may influence the final cost of construction. In many developing countries where there is political instability and volatile inflation rates, price fluctuation is a major challenge in construction cost management and improvement. In Chapter 9, Sections 9.2.1 and 9.2.2, market research is discussed as well as local alternatives and strategies for the continuous improvement of price fluctuations in basic construction projects.

a A library construction project was contracted to an experienced contractor with other subcontractors specialising in university building construction. The project is supposed to last for 18 months, and milestone lump sum payments have been agreed upon. In the eighth month of the construction project, the subcontractor complained of an increment in the prices of gravel and cement. Price fluctuations were not factored into the construction cost, and the bill of quantities did not include changes in construction prices. The contractor's quantity surveyor assessed this increment to include an additional 15% on the construction cost. The contractor is considering the use of off-site construction whereby panel walls and precast lintels will be produced on the construction site to reduce labour and construction material costs. However, this was not the original arrangement on the drawings, construction method statement, and quantities bill. What steps should the contractor take to effect this change? Do you think the contractor is right to adopt a new construction method in the middle of the project to keep the cost within limits?

_____

_____

_____

_____

b In the same construction project as above, what steps should the contractor take to discuss the changes price of construction materials with the employer? In the ninth month of the construction project, the contractor further received news from the subcontractor that the country's economic situation has caused an additional increment of 10% in the cost of steel rods, electrical cables, and paints. The fluctuations in prices seem to have affected all construction materials within a short while, and there is no indication of a future reduction. Are there continuous improvement strategies that can be applied to ensure that this project is completed within the required budget? What are the implications of not doing anything about the project's prevailing price fluctuations?

_____

_____

_____

_____

12.3 Case 14: Changes in laws and regulations

Changes in laws before the commencement of a construction project must be addressed regarding their implications for the construction project. Examples of changes in labour and immigration laws may affect the availability of migrant construction workers and professionals for a project. Professional regulations on construction drive construction businesses, innovation, training, inclusion, equality, and diversity.

The UK government construction strategy was issued to reduce the construction cost by 15% to 20% in all public construction projects by using BIM from April 2016. Therefore, all new private finance initiative projects and associated public construction projects in the UK are compelled to use BIM. In a medium-scale construction company registered as a tier 1 contractor with the UK government, BIM training has begun. However, there is a challenge of expertise availability in 5D BIM because of the BREXIT rules on construction businesses that discourage skilled migrant construction workers from emigrating to the UK. The top management is concerned about adopting BIM in their government contracts and their capabilities to meet the expectations of reducing construction costs with BIM. What steps should the management of this construction company take to ensure they reduce construction costs effectively with BIM and recruit the required talent?

_____

_____

_____

_____

12.4 Case 15: Compliance with laws

The same construction organisation is struggling with the adoption of BIM in their government contracts. A BIM common data environment was created to share files during the construction of a prison. The drawings were initially produced in drawing (dwg.) format. However, recent developments in BIM have made it clear that BIM collaboration format classes (BCF) document format is a standard format. IFC formats are universally accepted and compatible with all computer-aided drawing and measurement applications. Although this will not influence the construction cost, the architect has been asked to produce the drawing in BCF format. This architect is still new to BCF format and will need the services of an expert to produce this new format. The cost implications of this change will go a long way in increasing the construction cost, and the construction company has realised that they need to invest in training. The company has decided not to outsource the production of the BCF drawings but rather provide training sessions for all workers to upgrade their skills. Do you think the construction company has made the right decision? Is training within the workplace an important continuous improvement strategy and why?

_____

_____

_____

_____

12.5 Case 16: Constructability or buildability check

Constructability assessment was discussed in Section 9.4.3 of Chapter 9. A constructability or buildability check is integral to implementing continuous cost improvement. Constructability checks involve capital savings, construction materials' quantification, accessibility, labour requirement, and access to plants and equipment.

a An architect working for an employer was asked to investigate the feasibility of executing the design on the site. The design contains circular sections which will require the expertise of skilled masons. The cost of skilled labour was assessed. The output of the constructability indicated that it would be more expensive to implement the design, and considerable changes must be made to source the labour and construction materials locally as part of the new changes in design. Do you think this strategy will reduce construction cost significantly?

_____

_____

_____

b The continuous improvement strategy of a constructability check is also supposed to reduce construction costs before the commencement of a construction project. A constructability check was carried out. However, during the construction project, the construction manager realised that the plant and equipment required to construct the underground drains, which is inclusive of utility holes, are not available locally. The contractor has decided to subcontract the underground drainage development. After carefully reviewing the constructability report, they found that the assessor had not addressed the underground services and electrical installations in detail. In future constructability assessments, what minor details should the construction manager ensure are included in the report?

―――――――――――――――――――――――――――――――――――

―――――――――――――――――――――――――――――――――――

―――――――――――――――――――――――――――――――――――

―――――――――――――――――――――――――――――――――――

12.6 Case 17: Design economics norms check

Design economics deals with design implications for the cost of construction. Concepts such as designing to cost involve early cost advice to inform the design and cost planning teams of the cost implications of their decisions on a construction project. In Chapter 6, Section 6.3.5, the implications of economic decisions on overhead cost and continuous improvement were discussed.

a   A City Council office has contacted their property department to evaluate the economic implications of designing a town hall, including several rooms, a store, and three office spaces. The gross internal floor area for this design is 342 m², and the project's final cost will be over £300,000 when considering the cost/m² of a similar rebased construction in the district. This cost is beyond the budget of the City Council. How should the property department approach the construction cost calculation and provide very good early cost advice? Do you think this is a good continuous cost improvement strategy?

―――――――――――――――――――――――――――――――――――

―――――――――――――――――――――――――――――――――――

―――――――――――――――――――――――――――――――――――

―――――――――――――――――――――――――――――――――――

b   The order of cost estimates was used to produce a more detailed assessment of construction cost as part of a design economic evaluation. The employer was advised based on this output. Within the order of cost estimates, value-added tax (VAT), fluctuations in construction materials prices, and inflation percentages are included. In this order of cost estimate, the quantity surveyor has underestimated the percentage of inflation and fluctuations of construction prices. This led to a lower estimated contract sum. What are the implications of this underestimation on the progress of the construction design and future cost plans? Should there have been a continuous cost improvement strategy to ensure that the right cost information was applied in the order of cost estimates?

―――――――――――――――――――――――――――――――――――

―――――――――――――――――――――――――――――――――――

―――――――――――――――――――――――――――――――――――

―――――――――――――――――――――――――――――――――――

12.7 Case 18: Cost norms check

Cost checking for errors is a continuous cost improvement technique that the contractor's quantity surveyor must apply. A cost norm check deals with the integrity of the cost information provided in the cost plans and the bills of quantities. This technique relates to the ideals of constructability, market research, and local materials in Chapter 9.

a A contractor has submitted a bid for a contractor that is inclusive of a priced bill of quantities. Cost checks were performed on the individual sections of the priced bills of quantities, and errors were not detected. The outcome of the tender analysis indicated that the employer's consultants provided a review inclusive of an extra amount in the rate build-up for the excavations and fillings. This led to overbudgeting and subsequent rejection of the bid. Subsequently, the contractor informed the quantity surveyor of this feedback. Upon a second review, the contractor used unreleased construction rates from another city and did not consider the location of the intended construction project. Rebasing previously used rates from another location is converting previous figures with time and location indices. How will rebasing old cost values affect cost checks and final construction costs? How can rebasing fit the ideology of continuous cost improvement?

_____

_____

_____

_____

b Likewise, the feedback from the tender evaluation indicated that the proposed construction project would be in a foreign country, and the cost information has not considered the rising exchange rate. The contractor's quantity surveyor has made a mistake in producing the rates without consideration of the location and host country. What other lessons can the quantity surveyor learn from the cost checks they performed on the bills of quantities before submission? Which other continuous cost improvement strategy may be adopted to ensure that construction costs are not only suited for the construction project but also for the location?

_____

_____

_____

_____

12.8 Case 19: Wastage check

In Chapters 1, 5, 6, and 9, waste reduction was identified and discussed as the core basis of continuous improvement in construction cost management. There are physical and non-physical wastes in construction activities that border on productivity enhancements. In cases of continuous cost improvement, waste checks must be conducted to eliminate non-value-adding activities in the cost plans, schedules, and construction sites. The concept of circular economy in the construction industry, as discussed in Chapter 9, Section 9.4.1, is a beneficial strategy for waste reduction on construction sites.

a A Gantt chart was produced for the subcontractor to produce a retaining wall for an outdoor skating venue. The contractor and construction management have problems with the subcontractor's decision to import construction materials from outside the country when local alternatives are available. Similarly, construction material delivery and handling have been assessed to increase construction waste and unproductive activities on-site. The subcontractor will not heed the advice of the construction manager. What will the implications be of the subcontractor's unproductive decisions for the construction process? What steps should the construction manager take to ensure that the subcontractor makes use of the local alternatives?

_____

_____

_____

_____

b In this same retaining wall construction scenario, excavations and sand filling commenced early enough for the construction manager to realise a mistake in the programme schedule for the project. There are omissions in the schedule that must be included to ensure productive hours on the construction project and payment of labour wages. The construction manager wants to proceed with the activities on-site to ensure that they meet the construction project schedule without any delay. Considering the decision of the construction manager, in the long run, will delays lead to unproductive hours in the construction project? What will be the implications of overlooking this omission in the schedule?

_____

_____

_____

_____

12.9 Case 20: Alternative technology check

There are innovative construction technologies capable of reducing construction costs continually throughout the lifecycle of a project. Alternative technologies are important in the application of modern methods of construction, as discussed in Chapter 5. Changes and updates in technology must be checked before producing construction method statements or BIM execution plans.

a   A local government in a developing country has recently received a World Bank grant to construct low-cost social houses. Your construction company has proposed a design which will be produced to construct 100 new two-bedroom bungalows of 80 m². Recently, you read about 3D printing in the construction industry and have evaluated the cost of the 3D printing machine and concrete. Your assessment of the previous design and plans shows that the project will be completed in over 25 months at a higher cost. 3D printing will deliver all 100 houses within five months. However, you have not used 3D printing before, and it is still new to the country. There is no local expertise to operate the 3D printing machine. Do you still think this is a good idea to save costs and meet the housing demands of the local government? How will you circumvent the challenges of adopting 3D printing in the procurement process?

_____

_____

_____

b   The construction of 100 houses with 3D will reduce on-site labour by 90%, and this will not provide jobs for the labourers in the local community. Your continuous improvement approach in reducing the construction cost has been criticised as counterproductive and non-inclusive of the local supply chain. The local politicians also criticised 3D printing and off-site construction as an alternative to a construction project in a developing country. How will you convince the local government council and politicians that the alternative technology and the idea of continuous improvement in construction cost management will save the government money? Also, how will you respond to the implications of reducing on-site labour and the inclusion of the local community in the construction project?

_____

_____

_____

_____

12.10 Case 21: Facility management issues

Continuous improvement is not only useful for the construction process but also for the extended life cycle of the property. If the average life span is 60 years, then specific costs will be incurred in keeping the property in the right shape for its function. This assessment can be made before construction. Facility management is different from the constructability assessment. Hence, as discussed in Chapter 9, Section 9.4.2, material life cycle costing elucidates guidelines for ensuring continuous cost improvement in maintaining properties.

a A stadium will be built for the next sports festival, and the life cycle assessment of the proposed sporting facility has indicated an annual incremental maintenance cost of over 80 years. The main elements contributing more annual cost to the life cycle cost of the stadium development include chairs that will seat 35,000 people and the natural turf for football matches that will require maintenance. The facility will not be used for sporting events until the sporting festival, which will be held every four years. The construction project and sports festival will create jobs for the local community. What is the importance of material life cycle costing assessment in the development of this stadium? Are there new modern construction methods that can be explored to reduce the cost of maintaining plastic chairs and the natural turf in the stadium?

---

b After a complete material life cycle costing assessment, specifications were produced fitting the project's local materials and labour. However, the stadium's maintenance will require the services of a specialist facility management company that is not locally available. The cost of outsourcing the facility management to a specialist takes up 35% of the life cycle cost. What alternative arrangement must be made to ensure that facility maintenance costs are reduced before the commencement of construction activities? Is this a continuous cost improvement technique?

12.11 Case 22: Long-term maintenance agreement drafts

Long-term maintenance agreement drafts are very similar to facility management issues of material life cycle costing and providing good alternatives to construction components that will lead to higher life cycle costs. Section 12.11 presented the case of a stadium requiring the servicing of a leading specialist facility management company. If long-term maintenance agreements are drafted as part of the construction process, there is a need to drive down maintenance costs. In a similar construction project, the contractor involved in the stadium's construction was also asked to operate and maintain the facility for the next 30 years. You have been asked to review the contractor's proposal and draft a 30-year maintenance agreement. Based on your experience with continuous improvement, compile a list of focal areas requiring continual cost reduction strategies to maintain the stadium over the next 30 years. Do you think it will be beneficial to the employer to create a long-term agreement with the contractor who will construct the stadium? What continuous cost improvement lessons can you learn from previous large-scale construction projects with such long-term maintenance agreements?

_____

_____

_____

_____

12.12 Case 23: Electric power and water consumption check

In megaproject construction projects, especially construction over water, in the form of bridge construction, preliminary items constituting a large proportion of construction costs are important. Most preliminary costs are not quantified but prorated as a percentage of activities that cannot be measured in the contract sum. Preliminary construction items such as the provision of electricity, access to the site, access to water supply, scaffolding, insurance, site offices, and others contribute over 30% to the contract sum. Similarly, the cost of electricity and access to a water supply is consistent with cost attributes of preliminary item costs in construction projects.

a The general contractor (GC) has possessed a construction-site for the development in a very remote location with no electricity access. Before the possession of the site, the construction manager and quantity surveyor working with the GC omitted the access to electricity that will be necessary to power the concrete plant, site offices, security cameras, lighting for the construction site, and other related uses. You have suggested using a 300 KVA generator; however, the cost of diesel to fuel the generator throughout the construction process for 20 months will lead to an additional 25% cost to the contract sum. You have also identified additional costs in servicing the generator as part of the new preliminary cost. In this situation, will you buy or hire a 300 KVA generator? Do you think the capacity of this generator will be suitable for the construction activities, or would a lower capacity lead to a reduced cost of maintenance? Are there renewable alternatives to electricity supply on the construction site?

_____

_____

_____

_____

b There is no access to water supply in this same construction project because it is in a remote area. Based on your experience, you suggested sinking a borehole to provide a stable water supply for all construction activities. There are regulations regarding drilling boreholes in this district, and approval must be sought before you can provide a borehole. The provision of a borehole will delay the construction process by a month or more. Water supply is essential on the site, and your contracting company is ready to ensure they do all they can to provide a source of water supply. In your preliminary cost estimate, you have also found that you need to subcontract the drilling of the borehole to a specialist subcontractor. You have informed the contract administrator and employer of the delays in providing a water supply; schedule extensions have been granted under the contract conditions because access to the water supply was omitted in the unpriced bills of quantities provided to the employer. Do you think the contractor has made the right decision to inform the employer and contractor administrator? What other alternatives do you have to provide water on the site?

12.13 Case 24: Production continuity

Production continuity can form a major part of the constructability assessment discussed in Chapter 9, Section 9.4.3. The concept of production continuity concerns workflow leading to the progression on the construction site. Thus, the construction materials, labour, plant and equipment, and financial flow to the contractor mainly contribute to production continuity on the construction site. Chapter 1 also discussed the principle of continuous improvement in cost management by enhancing productivity in construction workflow.

a The construction of a road cable-stayed bridge of 2.7 km and 160 m high towers experienced a setback when one member of the construction team on-site fell from 140 m while hoisting the second steel tower. The loss affected the morale of the remaining construction working on-site. As the construction manager of this project, you do not want to be unempathetic with your words and actions. However, construction activities on the site need to continue. After carefully reviewing the health and safety procedure that led to the death of this worker, you realise that all health and safety measures, including personal protective equipment (PPE), were in place. Investigations were conducted to ensure that such an occurrence will not repeat itself. However, it was difficult to identify what led to this fatal accident on the construction site. How will you engage with the rest of the construction team, including the subcontractor and suppliers sourced locally? What measures must you take to ensure the continuity of this megaproject?

_____

_____

_____

_____

b The construction manager investigates the causes of the delays and increasing cost of construction. The outcome of the investigation identified the unproductivity of the workers on-site as the leading factor. Although it was difficult to identify the group of construction workers that are being unproductive, the construction manager must ensure continuity of the construction project. Provide a clear strategy to improve the productivity of the construction workers using continuous improvement strategies.

Secondly, the interim evaluation of the construction project by the employer's quantity surveyor and the Gantt chart of the construction programme indicates a major challenge in delivering the project within the agreed schedule and budget. How will the unproductivity influence the agenda to reduce the cost of production throughout the phases of the project?

_____

_____

_____

_____

12.14 Summary

The 24 cases on continuous cost improvement attributes in the construction industry presented in Chapters 11 and 12 have postulated the need for a micro-level implementation of continuous improvement in cost management practices and ensuring continuity of the project schedule. Therefore, when trying to attain continuous cost improvement in construction projects, there is a need to deliberate on the interaction of cost with health and safety, facilities management, all preliminary items of work which are peculiar to the construction site, specific cost variables for the project, and changes in law and technology. Waste cutbacks on construction projects depend on productivity and quality enhancement and continually eliminating non-value-adding activities.

13 Cases on Openness to Continuous Cost Improvement in Construction Organisations

Temitope Omotayo, Udayangani Kulatunga, and Bankole Awuzie

13.1 Introduction

The culture and behaviour of construction organisations dictate the type of strategy for innovation and the acceptance of continuous improvement. Chapter 7 gives a global perspective of organisational cultures of adhocracy, clan, market, and hierarchy on every continent. Openness to new ideas such as continuous cost improvement in construction organisations is essential in advancing project success, client satisfaction, and organisational competitiveness and growth. This chapter presents cases and examples from a survey of organisational culture behaviour in responding to innovative ideas.

13.2 Case study on openness to new ideas in construction organisations

A survey was conducted in 84 construction organisations and 135 respondents in Nigeria, where there is a combination of market and hierarchical cultures. The respondents in this survey were quantity surveyors and construction project managers involved with cost and project management. Openness to new ideas and innovation within these organisations was investigated to determine the organisation's perception towards change. If the employees feel their ideas cannot be passed on to the management, then continuous improvement cannot exist in such a workplace.

The responses are based on the perceptions of the employees rather than those of the management. This provided more suitable answers than the management would because the organisation may want to defend itself by providing positive responses in most cases. Nonetheless, it was determined that 9 respondents out of the 135 responses were not open to new ideas or innovation from the employees. Sixty-five respondents noted that their company was slightly open to new ideas. This implies that not all ideas are welcome, and the respondents find it very difficult to communicate

DOI: 10.1201/9781003176077-16

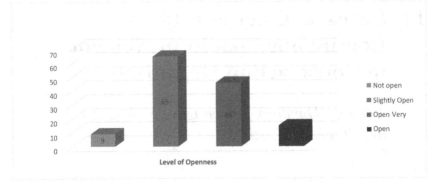

Figure 13.1 Bar chart showing the level of openness to new ideas.

Source: Authors.

suggestions to the upper management. Forty-six respondents highlighted that their organisations are open to new ideas whenever presented to them. In this instance, the employees (respondents) find it very easy to communicate their suggestions and ideas to the upper management and the management acts on these to improve (Figure 13.1).

Fifteen respondents only answered the "very open" category. This category of respondents indicated that their organisation made very good use of their ideas to improve and innovate quickly. In such an organisation, the management can have a research and development team. From the findings, very few small- and medium-scale construction organisations respond to new ideas, while the larger population of the respondents believe that their organisation is slightly open to new ideas and innovation.

The respondents' position (cost and project manager) is observed to be very different for each of the questions. Thirty-eight quantity surveyors in small- and medium-scale construction companies in Lagos, Nigeria, indicated that their organisations were slightly open to new ideas and innovation. Thirty-six quantity surveyors noted that their companies are open to new ideas and innovation, while only three quantity surveyors stated that their company was very open to new ideas and innovation. No quantity surveyor stated that his or her organisation is not open to new ideas and innovation.

The outcome of this survey is reflective of the nature of the market and hierarchical cultures where performance is dominant and fewer innovations are allowed. When organisations are completely open to new concepts and innovation, such as continuous improvement in construction cost management, opportunities for the market culture to change to an adhocracy culture will become possible. Organisational culture can change with the right leadership and external influence, such as government laws and regulatory bodies.

13.3 Case study on encouragement of continuous cost improvement

An organisation's ability to encourage new forms of cost control and management is an indication of accepting change. This is related to the organisation's behaviour towards continuous cost improvement, thus indicating the presence of an adhocracy organisational culture. Using the same case study in section 13.3, the respondents' views on this question reflected how well the organisation is willing to adopt continuous cost improvement. Although the management of an organisation can be unwilling to innovate and use the new ideas, the employees are ready to use continuous cost improvement. This was investigated, and the results indicate that over 70% of the respondents want a change in the construction cost management process adopted in construction projects. This is indicated by the number of responses, namely 95 out of the total population of 135. Only 40 respondents indicated that they did not want a change (about 30% of the respondents; Figure 13.2).

Fifty-five quantity surveyors indicated that they would encourage a new form of post-contract cost control, while an observed count of 22 noted that they would not encourage a new form of construction cost management. Forty project managers noted that would encourage a new form of construction cost management. Nevertheless, 18 project managers stated that they are satisfied with the post-contract cost control techniques used.

The respondents, who are mostly employees, are open to new ideas and innovations. Therefore, a problem with adopting new systems, methods, and techniques lies with the management of the companies. When it comes to implementing new concepts and innovation, the management function is crucial. In an adhocracy organisational culture, change is easier.

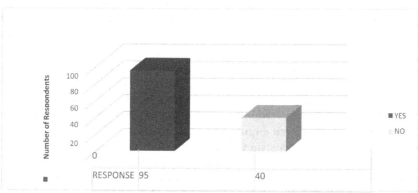

Figure 13.2 Bar chart illustrating the response to encouraging a new form of construction cost management.

Source: Authors.

The clan culture provides more difficulties for continuous cost improvement while the market and hierarchical cultures will still encourage a new form of construction cost management.

Cases under each organisational culture are discussed further to provide examples of how organisations behave under the scenarios of continuous cost improvement. The cases will look at the following:

- Clan culture and continuous cost improvement
- Hierarchy culture and continuous cost improvement
- Market culture and continuous cost improvement
- Adhocracy culture and continuous cost improvement

Hint for the cases

Continuous cost improvement is hinged on organisational behaviour, openness, and adapting to change for better competitive advantage. Continuous cost improvement must be implemented at the organisational level before the project execution.
Important reading: Chapters 7 and 8

13.4 Case study on clan culture and continuous cost improvement

Chapter 7, section 7.10.1 indicated that the clan organisational culture has extremes, and nepotism, and a family-like attitude towards improvement and performance. Countries such as France, the US, China, Saudi Arabia, Brazil, Canada, and India exhibit the clan culture. It was also noted that there might be multiple organisational cultures in a country. The clan organisational construction industry makes it very inflexible for continuous improvement to penetrate organisations. The following cases will present scenarios from a managerial perspective with emphasis on cost overrun and post-post reviews.

A construction organisation based in France has a staff capacity of 300 and has been involved in megaproject construction of dams and power plants. The megaprojects have always experienced cost overruns of over 40%. The construction establishment is mainly owned by a family that has a long history of civil engineering projects. In recent times, the poor performance of the construction organisation has caused severe financial problems for the owners. The company's competitiveness is at its lowest ebb, and suggestions are being made to sell the company or bring in more investors. The clan culture is an obstacle to innovation, and there has been ongoing nepotism for many years. Some of the professionals in the organisations with expertise in cost engineering have proposed embedding continuous improvement execution plans and reorganising the organisational culture to be more receptive to new trends such as

BIM, 3D printing, and continuous professional development for all staff, including the management. Do you think this is the right approach? If not, please suggest other approaches to change the existing closed clan culture in the organisation.

However, the construction projects team has a filing cabinet for the paperwork. Drawings, bills of quantities, and reports are usually printed out, and the office overheads are high.

Secondly, in a similar organisational culture in Brazil, there is open communication between the staff and the management team. One of the staff has suggested the application of digitisation of all existing documents and cloud storage. The management team has resisted this suggestion because they are used to the traditional pen, paper, and printers. Provide a better suggestion of how this staff should explain the benefits of digitisation to the older management team and what the implications of the continued use of pen, paper, and printer are for the organisation's operations.

Thirdly, in a similar situation, a construction organisation in China specialises in turnkey engineering projects. The organisation has produced several plant rooms for hotels, hostels, conference centres, and hospitals. Just as the Brazilian organisation has much paperwork, this Chinese construction company uses the same designs, and in recent times, the cost of steel has risen by 20%. The factory production system must improve to meet the rising cost of production materials, especially steel. Post-project reviews and the Toyota Production System, which made use of continuous improvement, have been suggested. A large amount of data and paperwork available will be very difficult to review. The software has been suggested for a post-project review of the production of the plant rooms and other similar projects. The first post-project review was conducted, and the recommendations were to replace some members of the design and cost engineering team after 15 years. This recommendation was made because there have been repetitive errors in the design and cost engineering processes for many years. This has led to severe financial losses in the organisation. The management team has rejected this recommendation and cited the relevance of the organisation's design and cost engineering team. It is suspected that this decision was made because of the relationship the affected staff have with the management team. What will be the implications of this decision made by the management team members for the organisation?

How can you effectively manage a clan culture with continuous cost improvement practices?

13.5 Case study on hierarchy culture and continuous cost improvement

Section 7.10.2 from Chapter 7 discussed the hierarchy of organisational culture in construction companies. A hierarchy culture focusses on individual and organisational performances. This type of culture is slightly open to new ideas that can drive performance, but it is mainly based on existing strategies and plans. The agenda of a hierarchical culture is profit, delivery time, quality, and more construction projects. There are also several extremes to a hierarchy culture in the construction industry, as discussed in section 7.10.2. India, Chile, Argentina, New Zealand, and Nigeria were identified as countries with the attributes of a hierarchy culture in construction companies. The examples in this section will provide more understanding of how a hierarchical organisational culture can clash with the ideals of continuous cost improvement.

a A prominent construction company in India founded in the 1920s specialises in engineering, construction, and real estate development. Over the years, this construction company has also been involved in water and industrial projects. The strong management team, which has its core values in competency and well-defined key performance indicators (KPIs), is looking to make more profit in constructing residential housing as part of its real estate expansion plan. The challenge facing the organisation is the pressures placed on their staff and the need to innovate. The commercial managers have been given financial targets for the year, and they need to propose real estate projects that will meet the organisation's financial targets. Continuous training within the workplace has been suggested by one of the commercial managers. This commercial manager wants to focus on continuous training of the design and cost management units. This approach will allow all the essential staff to keep themselves up to date with the latest development, technology, and design and cost management practices. Do you think this approach will create more adhocracy in a hierarchical organisational culture in the company?

----

b A New Zealand-based construction company is the leading contractor for public and commercial buildings. In recent times, the company has been inundated with challenges of cost overruns in their commercial project. Their main client is the government, and there have been concerns that their project delivery performance practices are not meeting the client's expectations. Continued cost overruns in their existing public building projects may lead to a strained relationship with the client. The commercial director is working on alternative solutions to reducing the occurrence and impact of cost overruns on their uncompleted projects. Some of the proposed strategies to alleviate incidents of cost overrun in this hierarchical organisation are creating a cost control department, managing risk issues that have an impact on the cost, having frequent reviews and giving feedback while the project is in progress, and making adaptations while the project is still in progress. This should help in applying the lessons learnt at the early stages of a particular project to the next remaining parts of the same project. The changes in the organisation's structure will be flagged by the managing director and will not be implemented because of the annual strategic framework. Suggest alternative ways of creating more flexibility in this construction organisation to meet the proposed changes in culture and construction cost management processes.

_____

_____

_____

_____

c The management of the Chilean construction company has over 200 full-time staff. The company has multiple public and private construction projects across the country. The staff have decided to stop working in their offices and on the construction site. They have also staged a protest in front of the headquarters. They have complained of a hostile working environment, lack of promotion, and difficulties in meeting the expectations of the construction company. If the strike does not end, the company will experience financial losses, and its reputation will be tarnished owing to the non-delivery of the project when due. As a board member of the management of this organisation, you have realised that the working environment may not be conducive to productive thinking and performance. You have raised this during the board meeting and emphasised the implications of the working environment on the cost performance of construction projects. The board disagrees with your stand and has recommended the termination of appointments of the staff who are union members. If the board decided to follow the path of the former, what would be the impact of their actions on the current construction project delivery? What should be the best course of action in this situation?

_____

_____

_____

_____

13.6 Case study on market culture and continuous cost improvement

Section 7.10.3 of Chapter 7 discussed the market organisational culture. In the market culture, specific external influences from the government and regulatory bodies may drive the innovation and acceptance of new ideas in a construction organisation. This is not entirely based on the willingness of the organisations, but market or economic forces and policy changes by the government will compel organisations exhibiting the market culture to innovate. The competitive advantage of an organisation with a market culture is paramount, and performances are rewarded. There is a level of open communication that can further implement continuous cost improvement in market culture-led organisations. The countries identified and reviewed as key market organisational cultures in Chapter 7 are Egypt, South Africa, the UK, and Nigeria. The examples of market culture and continuous cost improvement will consider inflation, government directive, and regulatory bodies' influence on project organisations.

a The UK finally left the European Union (EU) single market on January 31, 2020. Before this period, there had been fewer construction workers from the EU, and many constructions organisational in the UK had been forced to either employ migrant workers from outside the EU or adopt alternative construction methods to meet construction output demands. A construction company that acts as a supplier of granite, cement, bricks, cross-laminated timber, and steel is experiencing more difficulties in importing building materials from the EU because of Brexit. The cost of importing the building materials has increased owing to the tariffs. The company's management must respond to the market conditions in a way that will not affect their finances. The construction company has decided to embark on the production of brick and other building materials they can manufacture. This will save costs and increase their assets. Construction materials such as steel and cement will now be imported from China and Australia. What is your opinion on the behaviour of this construction company? Do you think they are right to have made these decisions because of Brexit?

_____

_____

_____

_____

b   The recent COVID-19 social distancing restrictions imposed by the government of Nigeria have not affected onsite construction activities, but managerial office work has confined the team to their homes. Most office-based construction professionals are not used to working from home, and there is no provision for Internet access. A construction company within Nigeria with a similar challenge has decided to respond to the challenges posed by COVID-19 by creating a communications web portal, which is secure and available to the staff. The management also provided Internet access to ensure that all staff can deliver their common construction tasks without using other publicly available means. How will this move influence cost performance and delivery of ongoing construction projects? Do you think it was a waste of financial resources to provide the web-based portal and Internet access? Would you label the provisions of the company as continuous improvement strategies?

c   A South African-based prefabricating company specialises in panel walls and volumetric construction. Their main client is a property developer who constantly demands prefabricated panel walls. The design and production of the panel walls are based on a similar design, and the production cost of cement has increased owing to an inflation rate of 3%. The prefabrication company has already received the order to produce the concrete panel walls. The production of the panel walls will lead to a decreased profit at the current cost of production. Other competitors in the construction industry are ready to offer clients their product at a reduced cost even if the product does not lead to the expected profit. The management of the prefabricating company understands the competitive off-site production environment, and the company is ready to continue with the offer. How should this off-site construction company manage the impact of the increasing inflation rate on production activities? How can one adopt a continuous improvement strategy to mitigate the impact of inflation on production in a competitive environment?

13.7 Case study on adhocracy culture and continuous cost improvement

Adhocracy in the construction industry was identified in Germany and Australia. As discussed in section 7.10.4 of Chapter 7, the adhocracy organisational culture is characterised by the attributes of innovation, creativity, and encouragement of new ideas within an organisation. Continuous improvement principles in an organisation thrive easily in the adhocracy organisational culture. The case examples in this section will consider the latency of continuous improvement in construction organisations and how they pervade construction cost management.

a A German-based building contractor has been making use of lean construction in its general contracting activities. They have also ensured that the employees communicate freely and receive the required training on the latest development in the construction industry. They further ensure that the creative freedom of the employees is not stifled. It is expected that this organisation will have a higher competitive advantage and will not experience difficulties in the delivery of the construction. This is not the case as they have constantly been experiencing cost overruns in their construction projects. In a commercial office construction project in the city centre, the project is 40% complete, and a cost overrun of 15% has already been recorded after the interim valuation. The rising cost of construction on the site has been attributed to the rising cost of construction materials and the unfavourable exchange rate experienced during importation. The causations of cost overruns were experienced in similar projects, and it seems there have been no lessons learned from previous projects. What should the general contractor and the management of the organisation do to ensure that the repetitive causations of cost overruns do not have a further impact on the ongoing projects?

_____

_____

_____

_____

b A consultancy firm in Australia is a BIM-led organisation involved in design and project planning activities. This organisation has an open communication policy in the firm to ensure that the employees share ideas and new concepts to further the company's competitive advantage. The adhocracy culture in the BIM-led organisation makes them meticulous and able to identify and eliminate errors in their design. The BIM-led design firm collaborates with a private quantity surveying firm to produce the bills of quantities for their 3D designs. The private quantity surveying firm does not have the same organisational culture as the BIM-led design firm. A bill of quantities was produced for a client after the 3D designs had been given to the private quantity surveying firm. This bill of quantities was for tender submission. Upon reviewing the tender and the bill of quantities, the reviewers identified several errors in the rate and cost components. The client has blamed the design team for the errors.

It was found that the private quantity surveying firm does not use BIM for measurement, and the traditional approach of making use of Excel or dimension papers is still being applied. How should the BIM-led design firm influence the culture of their partner quantity surveying firm with their adhocracy ideologies? Would you suggest that the BIM design firm end its partnership with the private quantity surveying firm?

c This same BIM-led design firm in Australia is working on designing timber frame prefabrication housing for a client who will then forward the design to the prefabrication company. The prefabrication company is based on the hierarchy culture where performance and output are more important than innovation. After the delivery of the prefabricated timber frame, the sizes did not fit the structure which was designed. The errors could be on the part of the prefabrication company and not the BIM-led design firm. However, the client does not view it this way because there will be an additional cost of construction resulting from this error. The BIM-led firm has been contacted regarding this error, and they have allotted the blame to the prefabrication company. In this instance, how should the BIM-led design firm advise the client and manage the relationship between the major stakeholders?

13.8 Summary

The case examples on clan culture innovation and change are difficult to accept by the organisation's management. The problems of nepotism and sentimentalism can stifle the ideals of continuous improvement in the workplace and construction management. In this instance, there is a need for a continuous improvement champion to explain the merits of the change in financial terms and further stress the implications of non-compliance to change. A hierarchy culture is based on performance, and the relationship between staff and the management team must be encouraged. A balance between performance and the encouragement of open communication in a hierarchical organisation should improve the culture and performance. The market culture is responsive to the industry's competitiveness and should always exploit continuous improvement practices within the organisation and in project delivery. The economic vagaries that may affect construction costs should be planned for with a financial risk management plan. In an adhocracy culture, continuous cost improvement already exists. However, when working with other organisations with a different organisational culture, the inter-organisational relationship should be based on their progressive culture. Diplomacy and knowledge sharing can rub off on partner organisational cultures that are resistant to change.

14 Cases on Continuous Cost Improvement in Procurement Strategies

Temitope Omotayo, Udayangani Kulatunga, and Bankole Awuzie

14.1 Introduction

Alharthi, Soetanto, and Edum-Fotwe (2014) postulated that construction procurement began in ancient Syrian and Greek cultures. Procurement in the construction industry is the strategic approach to acquiring services, materials, and supplies required to deliver built assets. It should not be confused with a supply chain of construction services and materials but the overall process involved in acquiring construction services such as architectural drawings, bills of quantities, and other contract documentation. Built asset products such as roads, bridges, residential and commercial buildings, dams, power plants, and other construction outputs must go through a procurement process. Hence, procurement strategies in the construction industry provide a holistic system of governing every process leading to the delivery of a built asset. This starts from the briefing stage of construction to the design, cost planning, scheduling, the application of BIM processes, contract documentation, tendering, award and negotiation of the contract, construction activities on-site, project evaluation, handing over, and the in-use phase of the built asset.

Continuous improvement in procurement strategies is more of a useful tool for enhancing all construction processes. In construction procurement, continuous cost improvement will behave differently under different strategies. This chapter presents the various cases under the procurement strategies under the sections of separated (which is also known as traditional), integrated (design and build variants), management, and collaborative approaches. Separate procurement routes feature sequential traditional and accelerated procurement routes.

The word "separated" describes the procurement process whereby each construction activity, such as design, bill of quantities preparation, and a programme of works, are separated from each other. For instance, the design must be completed before the bills of quantities will be produced, then the programme of works will be produced (Al-Harthi, Soetanto, & Edum-Fotwe, 2014). The *"integrated"* procurement approach does not separate the responsibility of design from construction but

DOI: 10.1201/9781003176077-17

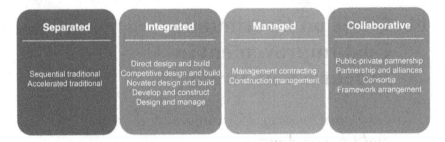

Figure 14.1 Categories of procurement strategies.

Source: Authors.

ensures that the designers are part of the construction process. The *"management"* makes use of a system of ensuring the contractor acts as a manager. The collaborative procurement category consists of an array of procurement strategies based on the ideals of long and multi-term procurement frameworks. Hence, the collaborative procurement approach ensures that all parties involved in the construction process are part of a charter to either continually for the client continually or manage construction projects delivered for many years. Further explanations of the procurement strategies in Figure 14.1 under the four categories will be given in Section 14.2.

The individual procurement strategies illustrated in Figure 14.1 for all four categories will be used to discuss different cases in construction procurement and continuous cost improvement. Continuous cost improvement applications in procurement strategy will feature the following cases:

- Cases on sequential traditional
- Cases on accelerated traditional
- Cases on direct design and build
- Cases on competitive design and build
- Cases on novated design and build
- Cases on developing and construct
- Cases on design and manage
- Cases on management contracting
- Cases on construction management
- Cases on public–private partnership
- Cases on partnerships, alliancing, and consortia

Unlike other case study chapters, there are no chapters to refer to when explaining the key procurement concepts. Therefore, each procurement strategy will be explained before the cases are presented in each section.

> **Hint for the cases**
>
> *Continuous cost improvement in procurement strategies includes PDCA that aims to lessen the demerits of the strategy on cost, quality, schedule, risk, and other identifiable deliverables of a construction project.*
>   Important reading: The brief background of each procurement strategy starting from Section 14.2–14.13. Read Chapter 9.

14.2 Cases on sequential traditional procurement

The sequential traditional procurement route is a traditional procurement system that is widely used around the world. A sequential traditional procurement route is a design-bid-build process that involves several consultants and teams who work independently (Gan, 2011). This procurement strategy describes a system whereby the designer, who is usually a consultant, provides the construction project drawings and other consultants work separately (Alharthi et al., 2014). The appointment of the contractor may also be conducted in conjunction with the team. The overall process of contract document and tendering before construction activities is a very long process. A sequential traditional procurement strategy is synonymous with price and cost certainty. Bills of quantities are completed before the tendering process and award of the contract. One of the challenges facing the sequential traditional procurement route is the time required to produce all the contract documents. The quality of cost certainty depends on the quality of outputs, such as the cost plans and bills of quantities derived from the cost planning process. The cases below consider continuous improvement in the cost management of a sequential traditional procurement strategy.

 a A dentist has decided to establish a private practice and has recently contacted you to brief you about what is expected of proposed dentistry construction project. After careful consultations, you have considered the traditional sequential approach because of its ability to provide cost certainty to meet the client's requirements. The client has also considered a quality requirement and is not bothered with the delivery time; however, the client has cost limitations that must be taken into consideration. As an expert in the continuous improvement process using the PDCA, you have suggested that the quantity surveyor adopt a continual cost reduction mechanism and design and cost approach. The quantity surveyor works independently from your practice and does not understand the implications of PDCA or designing to cost in the traditional procurement route. Please provide a brief explanation of PDCA and designing to cost as a tool for continuous cost improvement.

b   During the execution phase of the dentistry building, a contractor who was selected using the open tendering process is behind schedule. The client has contacted you to complain about the unexpected delays. Previously, the client had appointed you as the architect/contract administrator involved in the design and contract management. Based on the brief, the schedule was not important, and you have not prioritised it. This was why you did not choose any other type of procurement route to fast-track the process. Although the construction cost is still within budget, delays in a project may inadvertently lead to an overrun. To prevent an overrun proactively, which continuous cost improvement approaches would you advise the contractor to adopt in order to reduce the schedule and eliminate the risk of a cost overrun?

14.3 Cases on accelerated traditional procurement

An accelerated traditional procurement strategy is a sequential traditional procurement strategy whereby partial designs are used as the main contract document to select the contractor and proceed to the construction stage (Cheung, Lam, & Leung, 2001; Gan, 2011). An accelerated procurement strategy is a fast-track variation of a sequential traditional strategy. However, the early appointment of contractors with limited design information will lead to less cost certainty and cost overruns if the construction cost process is not properly managed. Accelerated procurement also involves numerous consultants and more financial risks in the execution phase. Continuous cost improvement can be realised in accelerated procurement with the implementation of the PDCA principle and the strategies elucidated in Chapter 9.

a   An employer has decided to use the accelerated procurement method to construct a new university library. The architect in charge has decided to involve the contractor very early after the conceptual design phase under the two-stage tendering arrangement. The architect is conversant with continuous improvement and has decided to adopt it in design, cost planning, and risk management. A pre-construction service agreement was signed in the first stage, and the second stage of the contractual arrangement under accelerated procurement involved negotiating the contractor's price. Under this arrangement, nearly all the risks were transferred to the contractor after the negotiation. A risk register was created, and the risk ownership column was created to allocate most of the risks to the contractor. In the process of construction, the contractor appointed a subcontractor with the permission of the architect. However, the contractor failed to transfer the responsibility of the risk register to the subcontractor, and most of the risks occurred. The contractor will take responsibility for the risks, which will not lead to additional costs. In this situation, which continuous cost improvement strategy should have been adopted by the contractor to ensure that risks management is transferred to the subcontractor and there is no impact on the project finances?

_____

_____

_____

_____

b In this accelerated procurement approach, construction cost is very difficult to determine at the early stages because of incomplete drawings. Hence, a bill of approximate quantities was produced to provide an idea of the construction cost. It was very difficult for the employer to know the exact construction cost for certain and continuous cost improvement to ensure that the cost is kept within an agreed benchmark. Re-measurement of the construction cost was conducted after each milestone completion on-site. How can you ensure the continuous cost improvement identifies waste factors before and after each milestone completion?

_____

_____

_____

_____

14.4 Cases on direct design and build

Under the integrated procurement arrangement, "design and build" is a key nomenclature. However, the term is only important in describing the broad category of procurement under the integrated approach. Direct design and build is the most common type of integrated procurement category. Direct design and build, also known as a packaged deal procurement, implies that the contractor has the sole responsibility to design and build the project (Boudjabeur, 1997). The contractor selection is faster than the separated procurement strategy because of the early involvement of the contractor just after the briefing phase. In contractor selection under direct design and build, the contractor can produce a proposed design as part of the tendering process. Contractors are appointed to take most of the risks from the employer, and the final cost of construction is unknown until project completion (Cheung et al., 2001; Alharthi et al., 2014).) This makes it more difficult for the cost to be controlled when compared to traditional sequential procurement. The essence of continuous cost improvement indirect design and build is risk and cost escalation minimisation.

a A direct design and build approach was justified for constructing new student accommodation that must be handed over to the client before the September resumption date to accommodate university students. The cost limits were set for the construction project, and the contractor has decided to bear all associated risks associated with design and construction. The contractor has identified escalated costs resulting from the transfer of a similar design used in a previous project. The cost of construction materials has increased, and there have been new constructability constraints that will lead to higher costs. Which of the continuous cost improvement strategies will you encourage the contractor to adopt?

_____

_____

_____

_____

b In future construction projects where design and build will be adopted, how will you explain to a contractor that the previous designs must undergo continuous improvement measures because of approximate cost estimations? In most instances where continuous cost improvement is adopted, older designs can be used. However, when older designs are used by the contractor, it is essential to improve upon them before application. Similarly, how can you use continuous cost improvement to enhance existing approximate estimates that will be reused in similar designs?

_____

_____

_____

_____

14.5 Cases on competitive design and build

Competitive design and build is a new variation of straightforward design and build. Take note of the nature of contractor selection and the demerits of direct design and build. Most contractors are selected very early in the direct design and build with limited competition, and there is limited cost certainty, whereas competitive design and build merges the features of a sequential traditional procurement strategy with design and build by ensuring that the contractors are selected based on competitive tendering (Boudjabeur, 1997). Competitive design and build provides an opportunity for continuous cost improvement in the cost management process of construction.

a Government employers have asked for proposals for a new bridge which will serve as an alternative to the existing one. As part of the competitive design and build procurement process, the contractors are expected to engage in a competitive proposed design, cost, schedules, construction method, and resources availability. The employer has chosen the best contractor with the right experience and access to resources. The cost of the construction project is also known. In this procurement strategy, the risk will be shared between the contractor and the employer. However, the construction delivery is at a slower pace when compared to direct design and build. There has also been an increasing delay in the delivery of construction materials from China. Further delays will lead to cost overrun. Suggest a continuous cost improvement strategy to alleviate the impact of the delays on the construction process.

_____

_____

_____

_____

b Upon delivery of the construction project, the inspection team has noticed lower quality materials were used in the road paving. An enquiry into the quality of the road paving revealed that the contractor has been unethical in delivering the project and has not followed the proposed design to the letter. The contractor has been asked to conduct remedial work on the paving. Do you think this procurement strategy may encourage contractors to cut corners in the process of enshrining continuous cost improvement?

_____

_____

_____

_____

14.6 Cases on novated design and build

Novated design and build comprises a variety of integrated procurements whereby the employer transfers the architect and another consultant team responsible for producing the detailed drawings, bill of quantities, a programme of works, and other tender documents to the contractor's team (Boudjabeur, 1997). Therefore, in this process, the consultant team will work on behalf of the employer and the contractor. In this scenario, once the contractor has been appointed, the consultant's payments are made through the contractor. Apart from the benefits of ensuring high-quality project delivery, trust issues may occur on the construction site because the contractor may feel monitored. The activities of the contractor on the site may also conflict with the agenda of the architect and other consultant teams, especially the quantity surveyor. Continuous cost improvement practices may be difficult to adopt in a novated design and build project when the contractor or novated consultants have no inclination to improve concepts.

a A novated consultant team working with the contractor had previously exploited the concepts of continuous cost improvement in a recently completed project. The team had also conducted post-project reviews and identified health and safety implications resulting from the social restrictions in the country. The novated consultant team has advised the contractor to adopt some new health and safety measures to address the social restriction concerns to reduce construction costs. The contractor sees no reasonableness in the suggestions and has rejected the recommendation. The consultant team then decided to report the contractor to the employer. How should the consultant team have reacted to the rejection of continuous cost improvement on the project? Was it right for the consultant team to report the contractor to the employer? What do you think will be the implications of the novated consultant team's action?

_____

_____

_____

_____

b The consultant team novated to the contractor has now realised a clash of interest on the construction project, which may lead to a dispute, delays, or increased construction cost. The contractor is unhappy with the activities of the consultant team and has decided to conceal some information from them by making use of another team. How should the novated consultant team approach this challenge with their experience of continuous cost improvement? Do you think it will be easier for the novated consultant team to organise a meeting which will lead to further training for continuous cost improvement on the project?

_____

_____

_____

_____

14.7 Cases on develop and construct

Develop and construct variations of an integrated procurement strategy is very similar to the novated design and build. The difference between develop and construct and novated design and build is that the contractor inherits the consultant team's incomplete design and further develops on the design by the contractor who also executes the project (Boudjabeur, 1997). Unlike novated design and build, direct design and build where the contractor may either not produce the design or provide the complete design, develop and construct affords the contractor some leverage on risk-sharing and working with a considerable design concept produced by the employer's architect. Hence, continuous improvement of design can be achieved with further direction from the brief, leading to further enhancement in the cost management process. In certain instances where the design which was inherited by the contractor has hidden errors, such errors may be transferred to the construction process.

a The architect advised an employer who intends to procure a private new build school on the develop and construct approach. The early involvement of a contractor led to further development of the conceptual design and construction. A bill of quantities was produced to ensure cost certainty, and the programme of works was subsequently produced. In the project's construction phase, the contractor realised that additional costs of external works will be incurred, and this error was derived from the inherited survey plan. Before reaching this phase of the construction project, which continuous cost improvement strategy could have been used during the design improvement phase?

_____

_____

_____

_____

b As the construction project progressed, more errors were identified in the design, which led to major variations. The contractor is now blaming the architect who produced the conceptual design and unclear information in brief. The cost implications of the major variations will lead to an additional 15%. Which of the construction cost improvement strategies can the contractor adopt to ensure the variations do not impact the construction project? Furthermore, what could have been done by the contractor to prevent errors in the design even though it was inherited as part of the construction project?

_____

_____

_____

_____

14.8 Cases on design and manage

Although the design and management procurement may be categorised under the management procurement approach, design and management are considered an integrated procurement because the contractor is employed to produce and manage the construction project (Boudjabeur, 1997; Cheung et al., 2001). The design and manage procurement route is an efficient and fast-track procurement approach that ensures that the contractor not only designs and constructs the built asset but also engages in future management. An example of design and management can be found in many oil and gas construction projects.

a A £5 billion offshore subsea construction project was designed by a specialised contractor to produce a subsea scope. At the end of this oil and gas project, over 450,000 barrels per day will be produced. The contractor was also responsible for managing the operations of this facility by providing schedule maintenance. A continuous cost improvement approach was used to produce the cost estimates required for this project. However, this ongoing project requires you to suggest continuous cost improvement measures that will keep this project within budget and ensure the life cycle costing is effectively planned.

_____

_____

_____

_____

b The subsea offshore project also uses a floating production storage and offloading (FPSO) mechanism to connect with the subsea sections in a design and manage a project that will last for five years. This megaproject is expected to be delivered within budget and schedule. However, there is a 1% overrun in the cost after 14 months. This overrun amount to £50 million extra costs. Which strategy would you adopt in this design and manage the project to limit the occurrence of overruns and ensure the project is delivered within the budget? In terms of a design and manage procurement approach, the leading contractor is a specialist who understands construction technology. How should this leading contractor adopt the concepts of continuous cost improvement to further the course of their professional development, including their staff?

_____

_____

_____

_____

14.9 Cases on management contracting

A management contracting procurement strategy is one of the fast-track management procurement options available in the construction industry. The management contracting approach uses a leading contractor who acts as a management contractor in charge of multiple works contractors (Boudjabeur, 1997; Alharthi et al., 2014). The specialist work contractors will then work on various project sectors, in most cases concurrently, to deliver each construction element to fast-track the construction process (Intaher & Johanna, 2012). The challenge with this procurement arrangement is conflicting interests among the work contractors who do not complete their activities on time. The scope of the work must be clear enough for the employment of work package contractors to deliver their portion of the project. Continuous cost improvement under management contracting may be complex to execute without a clear scope of work, a programme of works, and a construction method statement. The complexity of the construction project may also become difficult for cost control and management enhancements.

a An employer has been advised to use a management contracting approach to fast-track the delivery of a building associated with a train station. The management contractor has engaged the services of specialist work contractors for the excavations, filling, external works, masonry, building, electrical installations, plumbing, mechanical work, heating, and ventilation. The plumbing and electrical work contractors must conclude their activities before the tilers and painters begin their tasks. However, there have been delays in the installation of the electrical cables as some of the trunking materials have not been delivered. This delay will affect the tilers, painters, and other work-package contractors who should be on the site. How should the management contractor use principles and concepts associated with continuous cost improvement to ensure that this fast-track procurement approach does not lead to time and cost overruns? Please note that the management contractor has the direct obligation to ensure that the project is delivered within the expected time frame and budget.

b In a multi-storey car park construction, the management contracting procurement arrangement has been adopted to ensure that the project is delivered within 15 months. Other work package contractors were involved in the plumbing, electrical services, and elevator installations. The cost of concrete, cables, and PVC pipes has increased by 10 to 15% as the project progressed into the eighth month. How can the management contractor handle the escalating prices of construction materials under a management contracting strategy where completion time is as important as the cost?

14.10 Cases on construction management

Construction management procurement is very similar to management contracting. Construction management makes use of a construction manager who is appointed by the client as a consultant to advise the client on the construction process and choice of trade contractors (Boudjabeur, 1997; Intaher & Johanna, 2012). Professional service contracts will be signed between the consultant professional team and the client, and the construction manager and the client. In this procurement arrangement, the client is inexperienced in construction project management yet wants to be in charge of the construction process. Therefore, continuous cost improvement may become difficult or easier to implement in the construction process depending on the continuous cost improvement skills acquired by the client and construction manager.

a   In a multi-billion pound refinery construction project, the client's company has decided to use construction management to procure the building projects associated with the refinery. The in-house team from the property department has been contacted to supervise the building construction process. The construction management team is conversant with continuous cost improvement principles and has also employed trade contractors such as bricklayers, roofers, tilers, and builders. Under this arrangement, the trade contractors do not need to ensure that construction waste should be minimised, and the modern methods of construction that are being advised by the construction management team seem strange to the semi-skilled workers on-site. What should the construction management team do to upskill the workers on-site within a short time?

_____

_____

_____

_____

b   In another establishment with a property development team, continuous cost improvement is not important to the client in charge of the project, but the construction manager has advised the client on the PDCA practice for cost reduction and maintenance. The client is very experienced in traditional construction methods; however, modern construction methods, which are linked with continuous cost improvement, will imply learning new skills. The client is not ready to learn new skills that will aid the reduction of construction costs. How should the construction manager advise the client of the need for more training without causing any strain on the existing relationship in the organisation?

_____

_____

_____

_____

14.11 Cases on public–private partnerships

Collaborative procurement entails partnering between government and private organisations. Partnering as a collaborative terminology encourages a long-term framework for trust and openness between contracting parties who sign a charter to mutually engage in procurement activities for several years (Mohammad Hasanzadeh, Hosseinalipour, & Hafezi, 2014;

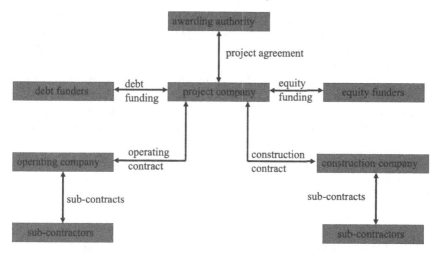

Figure 14.2 Private finance initiative (PFI) framework arrangement.
Source: Authors.

Hosseini, Wondimu, Bellini, Haugseth, Andersen, & Lædre, 2016). There is a dependency on each other for success and risk-sharing. In the UK, the private finance initiative (PFI) is a public-private partnership (PPP) created under a framework involving multiple parties, including the equity and debt funders (Carrillo et al., 2006). The quality of construction projects produced under the PFI is and other similar PPP arrangements worldwide have been a subject of research and debate. Hence, continuous cost improvement presents a challenge for PPP arrangements to improve the quality of construction outputs.

Under the PFI arrangement, as illustrated in Figure 14.2, several parties, including several subcontractors, are included in the framework. Continuous cost improvement is very difficult to include under the arrangement without authorising the awarding authority or project company.

a   The government has signed a 25-year agreement with a company to construct, manage and operate a hospital. The project company is faced with reducing construction overheads and sacrificing the quality of the construction process to meet profit expectations. Do you think the project charter and all stakeholders should be aware of the importance of continuous cost improvement? Where will continuous cost improvement fit into the framework of Figure 14.2? How can continuous cost improvement be adopted in the framework?

b In an ongoing PFI airport construction project, the construction company has adopted modern construction methods and continuous cost improvement. The project company is finding it difficult to measure the performance of the construction process under this arrangement. However, they have decided to focus on the cost performance of the construction project. Do you think continuous cost improvement can be used to monitor and maintain the cost of construction? What are the complexities involved in the use of continuous cost improvement under this PPP arrangement?

14.12 Cases on partnerships, alliancing, and consortia

Alliancing is a form of partnership framework arrangement whereby risk is shared among multiple parties in a contract for a larger-scale public–private project (Walker, Hampson, & Peters, 2002). Alliancing and consortia are forms of joint ventures which are common in the oil and gas industry. Alliancing and consortia are much riskier than PPP arrangements, and the dynamics involved in the relationships are more complex than any other partnership arrangement. Continuous cost improvement in the alliancing and consortia partnership must be based on mutual agreement between all parties.

a A national oil corporation has created an alliance with a multinational oil and gas company to explore crude oil for the next 20 years. Under this arrangement, refineries will be constructed, and multiple construction and oil and gas servicing companies will be working under a three-tier partnership arrangement. In one of the onshore facilities, the servicing company will have to main the installations every two years, and the cost of maintenance will increase annually. Is there a possibility of incorporating continuous cost improvement in this complex arrangement? What is the barrier facing continuous cost improvement in a more complex partnership arrangement such as this?

---

b The national oil corporation has instituted a training programme, and all parties involved in the consortia must be included in the programme. As part of the training programme, the principles of continuous cost improvement have been included. The benefits have been highlighted; however, most of the parties under the framework arrangement feel that continuous cost improvement will constitute additional cost and time while implementing it. Do you agree with this assertion? How can you integrate continuous cost improvement in a complex partnership arrangement to monitor and control construction cost and maintenance and operations cost?

_____

_____

_____

_____

14.13 Summary

Integrated procurement strategies involving sequential traditional are more detailed in cost planning, design, and tender documents. Continuous improvement fits the sequential traditional procurement approach more than other procurement options do because of the already established cost certainty in the project. In accelerated traditional and variations of design and build, continuous cost improvement strategies are likely to encounter a range of complexities relating to the contract structure, lack of cost certainty, and the need to produce more construction output more timeously. Similar consequences can be ascribed to management procurement approaches. Further complex challenges of implementing continuous cost improvement can be associated with PPP, partnership, alliancing, and consortia depending on the risk, the scale of the project, the time scale of the partnership framework, and the risk-sharing requirements.

References

Al-Harthi, A. A. S., Soetanto, R. and Edum-Fotwe, F. T. (2014). Revisiting client roles and capabilities in construction procurement. In *Proceedings of the International Conference on Construction in a Changing World – CIB W92 Procurement Systems,* Sri Lanka, p. 12.

Alharthi, A., Soetanto, R. and Edum-Fotwe, F. (2014). The changing role of the public client in construction procurement. In *Proceedings 30th Annual Association of Researchers in Construction Management Conference, ARCOM 2014,* pp. 403–412.

Boudjabeur, S. (1997). Design and build competition. *Electronics Education,* 1997(2), pp. 24–24. doi: 10.1049/ee.1997.0042

Carrillo, P. M. *et al.* (2006). A knowledge transfer framework: The PFI context. *Construction Management and Economics,* 24(10), pp. 1045–1056. doi: 10.1080/01446190600799224

Cheung, S. O., Lam, T. I. and Leung, M. Y. (2001). An analytical hierarchy process based procurement selection method. *Construction Management and Economics,* 19(4), pp. 427–437. doi: 10.1080/014461901300132401

Gan, S. (2011). Briefing: Specifics of construction procurement strategies in CIS. *Proceedings of Institution of Civil Engineers: Management, Procurement and Law,* 164(3), pp. 113–115. doi: 10.1680/mpal.10.00030

Hosseini, A., Wondimu, P. A., Bellini, A., Haugseth, N., Andersen, B., and Lædre, O. (2016). Project partnering in Norwegian construction industry. *Energy Procedia*, 96, pp. 241–252.

Intaher, M. A. and Johanna, A. B. W. (2012). Supply chain management challenges in the South African public sector. *African Journal of Business Management*, 6(44), pp. 11003–11014. doi: 10.5897/ajbm12.360

Mohammad Hasanzadeh, S., Hosseinalipour, M. and Hafezi, M. (2014). Collaborative procurement in construction projects performance measures, case study: Partnering in Iranian construction industry. *Procedia – Social and Behavioral Sciences*, 119, pp. 811–818. doi: https://doi.org/10.1016/j.sbspro.2014.03.091

Walker, D. H. T., Hampson, K. and Peters, R. (2002). Project alliancing vs project partnering: A case study of the Australian National Museum Project. *Supply Chain Management*, 7(2), pp. 83–91. doi: 10.1108/13598540210425830

15 Contractual Payment Arrangements and Continuous Improvement

Temitope Omotayo, Udayangani Kulatunga, and Bankole Awuzie

15.1 Introduction

Procurement practices in the construction industry are also supported by contractual arrangements based on standard forms of contract such as the Joint Contract Tribunal (JCT), the New Engineering Contract (NEC), the International Federation of Consulting Engineers (FIDIC), the Royal Institute of British Architect (RIBA) contracts, bespoke contracts, and other contract suites. In each of the contract suites, contractors and employers have a structured approach to payment for work completed on-site. Some of the arrangements are lump-sum contracts, fixed-price contracts, target pricing contracts, and others discussed in this chapter. The essence of payment arrangements in contract suites is to ensure that financial risks associated with a construction project are shared or transferred to specific parties in a contract (Davis, Love & Baccarini, 2008). Contractual arrangements are chosen regarding the nature of the project, employer's requirements, the type of procurement strategy, and standard forms of contract suite.

As indicated in Table 15.1, there is an array of options of contractual arrangements depending on the procurement route. For instance, NEC4 consists of the target pricing option, which will suit a sequential or accelerated traditional procurement strategy. Likewise, we have the fixed price and lump sum contracts within the JCT contract suite for traditional procurement options.

The effectiveness of construction cost improvement strategies in varying contractual arrangements behave differently under specific scenarios and projects. The attributes of the contractual arrangements are explained from Sections 15.2 to 15.8 under the discussion of the following cases:

- Cases on lump-sum contracts
- Cases on fixed-price contracts
- Cases on remeasurement contract

DOI: 10.1201/9781003176077-18

- Cases on cost reimbursable contracts (prime cost contracts or cost-plus contracts)
- Cases on guarantee maximum price
- Cases on unit pricing contracts
- Cases on target cost contracts

Table 15.1 Procurement strategies, standard forms of contract, and contractual arrangements

Procurement strategy: Sequential traditional	**Procurement strategy:** Accelerated traditional	**Procurement strategy:** Direct design and build
The standard form of contract: JCT suite, NEC4 suite, FIDIC, RIBA, bespoke contract, etc.	**The standard form of contract:** JCT suite, NEC4 suite, FIDIC, RIBA, bespoke contract, etc.	**The standard form of contract:** JCT suite, NEC4 suite, FIDIC, RIBA, bespoke contract, etc.
Contractual arrangement: Fixed price contract, targeting pricing, cost reimbursable contract	**Contractual arrangement:** Fixed price contract, target pricing, remeasurement contract, guarantee maximum price contract	**Contractual arrangement:** Remeasurement contract, cost reimbursable contract with a fixed fee, lump-sum contract
Procurement strategy: Develop and construct	**Procurement strategy:** Novated design and build	**Procurement strategy:** Management contracting
The standard form of contract: JCT suite, NEC4 suite, FIDIC, RIBA, bespoke contract, etc.	**The standard form of contract:** JCT suite, NEC4 suite, FIDIC, RIBA, bespoke contract, etc.	**The standard form of contract:** JCT suite, NEC4 suite, FIDIC, RIBA, bespoke contract, etc.
Contractual arrangement: Lump-sum contract, fixed-price contract, guarantee maximum price contract, target pricing contract	**Contractual arrangement:** Lump-sum contract, fixed-price contract, cost reimbursable contract	**Contractual arrangement:** Remeasurement contract, unit pricing contract, cost reimbursable contract, target pricing contract
Procurement strategy: Competitive design and build	**Procurement strategy:** Construction management	**Procurement strategy:** PPP/PFI, partnering, and alliancing arrangements
The standard form of contract: JCT suite, NEC4 suite, FIDIC, RIBA, bespoke contract, etc.	**The standard form of contract:** JCT suite, NEC4 suite, FIDIC, RIBA, bespoke contract, etc.	**The standard form of contract:** JCT suite, NEC4 suite, FIDIC
Contractual arrangement: Target pricing contract, unit pricing contract, guarantee maximum price contract, lump-sum contract	**Contractual arrangement:** Lump-sum contract, fixed price, unit pricing contract	**Contractual arrangement:** Lump-sum contract, fixed-price contract, guarantee maximum price, cost reimbursable contract

---

Hint for the cases

Continuous cost improvement in the contractual arrangement is more about payments, financial risks, gain and pain sharing. The PDCA and other continuous improvement principles are geared towards ensuring the satisfaction of all contractual parties.
Important reading: The brief background of each procurement strategy starting from Section 15.2 to 15.8. Read Chapters 8 and 9.

15.2 Cases on lump-sum contracts

The lump-sum contract is one of the commonly used contractual arrangements between contractors and employers in building and engineering projects. In a lump-sum contract, the client pays the contractor an agreed amount of money without breaking down for a completed work section. A contract sum is derived from a complete design that must be known for tendering purposes (Murahashi, 1987). Under this contractual arrangement, the client bears little or no risk because of the price certainty. An accelerated traditional procurement strategy can be associated with a lump-sum contract where the contractor must begin construction very early. One of the demerits of a lump-sum contract is that it is very difficult to integrate with construction variations because of the agreed lump sum (Zainordin, Abd Rahman, Sahamir & Khalid 2019). This may lead to greater risk-bearing for the contractor, who may not have much influence over the designs and errors transferred from the employer's contractor. Continuous cost improvement is very difficult to integrate with a lump-sum contract from the contractor's perspective. However, if the construction method statement and other associated conditions were driven by the need to produce a built asset with a modern construction method and cost management, the contractor would be forced to engage in continuous cost improvement practices.

a   In a beachfront resort construction project, a lump-sum contract has been agreed upon with the contractor. Some landfilling was required to ensure the right foundation strength was achieved for the multiple building projects. Additional variation costs were identified in the panel wall measurement resulting from defective designs and errors in the bill of quantities. This variation will amount to £23,000 if the contractor has been asked to bear the risks based on the lump-sum agreement and feedback from the contract administration. Which continuous cost improvement can the contractor adopt to ensure the variations have a minimum impact on their financial position on the project? Are there modern methods of construction that could have made the lump-sum contractual arrangement easier for the contractor?

---

b In this same lump contractual arrangement and project, the contractor has not received a lump-sum payment for the panel wall construction after addressing the variations. The milestone payment terms were agreed upon as part of the JCT 2016 standard building contracts with quantities. The contractor has been working with this employer for several years and intends to maintain the relationship. After carefully considering the employer's financial position, the contractor has decided not to suspend construction activities on the site or seek an alternative dispute resolution. Which continuous cost improvement measures can be adopted to reduce construction costs pending the employer's time in making the necessary payments? The contractor is bearing many risks under this payment arrangement. Can you suggest another contractual payment arrangement suitable for this type of project to the employer?

15.3 Cases on fixed-price contracts

Fixed price contractual arrangement is different from a lump-sum contract. However, there is a combination of terminologies in the construction industry in the form of a fixed price-lump sum contract, where the features of a fixed price and lump sum contract are merged to provide equality in construction risk sharing. The distinguishing factor of a fixed-price contract is that the risk is shared almost equally between the contractor and employer (Davis, Love & Baccarini, 2008; Zainordin et al., 2019). Unlike lump-sum contracts, where variations are difficult to manage, fixed-price contracts are more flexible to changes in scope and variations in construction projects. However, the price negotiations make the overall process slower than a lump-sum contract. Sequential and accelerated traditional procurement strategies are associated with fixed-price contracts. The flexible nature of fixed-price contracts makes continuous cost improvement easier to implement in the construction process. Continuous cost improvement under fixed-price contracts provides an opportunity for the contractor and other stakeholders involved in the construction process to change the course of the project dynamically for the best outcomes.

a A contractor was appointed very early under the accelerated procurement route in a residential building project. After this appointment, the contractor was involved in shaping the direction of the design. The design team produced the design, which is just more than the conceptual design phase. A fixed-price contract was signed between the contractor and the employer. Under this arrangement, the contractor will be responsible for 55% of the risks, and the ownership of the risk was identified in the risk management plan. The risk management plan and contingency plans included the financial implications of the risk on the project. Strategies for sharing the risk and promoting waste reduction and recycling construction materials were also added to the construction project documents. The fixed-price contract also included gain and pain sharing measures. Is this a clear example of continuous cost improvement in a construction project? Do you think the PDCA approach is easier to implement under the fixed price contract arrangement?

---

b In this same residential construction project, the contractor has included a percentage change in the risk if it occurs. Insurance and indemnities were also added to the construction contract documents. However, the contractor realised that the cost of preliminary items, especially electricity and water supply, increased in the 6th month of this 19-month construction project. The project's price is fixed, and this construction risk must be owned by the contractor even though it was not identified in the risk management plan. Continuous cost improvement may still be possible under these circumstances if the overhead costs resulting from the increment in the preliminary items are either replaced or reviewed. Which of the continuous cost improvement strategies must the contractor adopt to ensure that the project does not reduce the expected profit?

15.4 Cases on remeasurement contract

The remeasurement contract is associated with the direct design and build and variants of the integrated procurement strategies whereby the contractor is appointed very early and expected to design and build a project. Direct design and build usually follows a process whereby the contractor is expected to submit proposals containing designs, costs, activity schedule, and construction method statements. Approximate bills of quantities are normally used in this arrangement, and remeasurement contracts can be signed. The contractual arrangement for payment can follow the remeasuring construction activities on the site to get a full construction cost (Davis et al., 2008; Zainordin et al., 2019). The demerit of the remeasurement contract is that there is no cost certainty, and the price is not fixed. Hence, the employer will bear most of the financial risk. The employer is responsible for cost and time overrun incidents, and the contractor may not deliver the best quality construction output due to this extensive free will. As a fast-track construction process, variations and scope are flexible. A remeasurement contract can also be used in management contracting procurement arrangements where the construction scope is very flexible, and the contractor is expected to bear little risk. This flexibility in financial risks makes it easier for the contractor to implement continuous cost improvement but difficult for the employer to accept the continuous cost and construction scope changes.

a   The management contracting procurement strategy is being used for the construction of an international conference centre. The schedule and quality have been prioritised, and the contractor signed a remeasurement contract. The employer's quantity surveyor has been to the construction site several times to physically measure the construction work, review the invoices of construction materials purchased, and prepare a full cost forecast of the project before valuing the amount due to the contractor. In one of the remeasurement activities on-site, the employer's quantity surveyor has made a mistake in the remeasurement. The work package contractor recently received timber studs for the partition walls, but this was not included in the quantity surveyor's valuation. The amount due to the contractor and the work package contractor was incorrect, and the contract administrator has decided only to accept the cost information provided by the quantity surveyor. Do you think this contractual arrangement can stifle innovations such as continuous cost improvement in the construction process? Because it is important to receive the right amount of payment due for the month, how should the management contractor clarify this mistake?

b   The contractor also has another construction project with the remeasurement contractual arrangement for payment under a direct design and build strategy. This contractor has produced the design and used approximate quantities. The remeasurement by the employer's quantity surveyor identified some new construction methods which were different from what was proposed, and the cost of the construction project would be higher than the cost limit. The contractor has been using the PDCA and continuous cost improvement strategies to change the scope of construction to suit modern methods of construction and innovation. The employer is unhappy with these changes.

How should the contractor communicate the benefits of the changes to the employer's representatives and contract administrator to ensure that continuous cost improvement is implemented?

_____

_____

_____

_____

15.5 Cases on cost-reimbursable contracts (prime cost contracts or cost-plus contracts)

A cost-reimbursable contract is also known as a prime cost contract or cost-plus contract. This contractual arrangement only ensures that the employer pays for the actual cost of construction, invoices, and other expenses incurred due to construction activities (Nkuah, 2006; Picornell, Pellicer, Torres-Machi & Sutrisna, 2017). Cost-reimbursable contracts usually contain specific pre-negotiation agreements, including percentages of cost for construction materials and labour cost; this is important to ensure that the contractor's overhead costs and profits are added to the contractual arrangement. The variations of reimbursable cost contracts are as follows:

- Cost plus fixed percentage
- Cost plus fixed fee
- Cost plus with guaranteed maximum price contract
- Cost plus with guaranteed maximum price and bonus contract

Cost-reimbursable contracts are associated with construction procurements where the scope of works is unclear or unknown, and some cost limits must be provided for the contractor's guidance. Direct design and build, novated design and build, and management contracting are procurement approaches where cost-reimbursable contracts can be used because there is no contractor's price certainty at the commencement of the project, and there are high flexibilities for variations (Davis et al., 2008). Scope changes can be accommodated under the cost-reimbursable arrangement, and the employer bears most of the risk and benefits from overspending. This contractual arrangement can be very beneficial for contractors who are willing to innovate in construction projects. Hence, continuous cost improvement can be achieved in a cost-reimbursable contract.

a The bill of quantities was prepared for minor works for the renovation of the toilets in an airport. The percentage for prime cost was added to the description of the works, which will include the supply and fixing of new sanitary appliances and fixtures. The minor works contractor received the specifications for construction works but decided to use another cheaper specification to reduce the construction cost. At the end of the construction project, the employer felt unsatisfied with the quality of the sanitary appliances supplied and fixed in the renovation. Do you think continuous cost improvement made the contractor greedy? How will you describe the actions of the contractor?

_____

_____

_____

b A novated design and build construction project has the consultant team working with the contractor. The contractor has signed a cost-reimbursable contractual arrangement with the employer. However, this contactor wants to reduce overhead costs and gain more profit from the construction project. The contractor has been colluding with the subcontractors and suppliers to inflate the reports and invoices showing the cost of construction materials, labour rates, and other construction costs. The employer and the novated consultant team are unaware of this arrangement. The employer bears the risk of changes in construction cost, and the contractor's unethical practices are difficult to detect. Do you think continuous cost improvement ideologies and strategies can encourage the contractor to become unethical? What are the implications of the contractor's actions?

_____

_____

_____

_____

15.6 Cases on guaranteed maximum price

Boukendour and Bah (2001) noted that guaranteed maximum price contracts have the features of cost-reimbursable contractors and fixed-price contracts. Davis and Stevenson (2004) also stated that guarantee maximum price offers price certainty to the employer. The guaranteed price is agreed upon before the construction work commences, and the description of work to be completed is well defined in the bills of quantities. The contractors bear nearly all the associated risks in the project, and there is less flexibility for changes in the construction project. This contractual arrangement is mostly used in traditional sequential procurement and competitive design and build procurement strategies. Continuous cost improvement will be beneficial to the client if there are limited or minor variations. However, the contractor must adhere to the contents of the bills of quantities and guarantee maximum price arrangements to ensure timely delivery of the construction project.

a Guarantee maximum price was agreed upon before the commencement of a road construction project. The contractor did not envisage rock boulders in the excavations and additional groundwater which had to be disposed of. Although the clauses covering relevant events and matters were referred to by the contractor, the contractor will still incur additional preliminary items of work cost under the guaranteed maximum price arrangement. The contractor wants to adopt continuous cost improvement strategies to mitigate the additional preliminary costs incurred from extra excavations and water disposal. Identify the additional preliminary costs for the construction above-mentioned activities and suggest continuous cost improvement measures to reduce overhead costs.

_____

_____

_____

_____

b The contractor under the guaranteed maximum price contract wants to use a circular economy strategy to reduce construction costs. However, the construction cost as stated in the bill of quantities must be followed to avoid breaching the contractual arrangement. The construction project involves constructing a new tram line across the city, and the sequential traditional procurement strategy has been adopted. What are the opportunities available for the contractor to include a circular economy strategy for continuous cost improvement in this project without interfering with the contract clauses? Do you think the contractor must include any new construction approach or strategy under this contractual arrangement?

_____

_____

_____

_____

15.7 Cases on unit pricing contracts

A unit pricing contract offers a convenient payment arrangement for the employer for every work completed onsite (Nadel, 1991; Miranda, Bauer, Aldy & Podolsky, 1996). This contractual arrangement pays for the hourly rates of labour and materials purchased. Management contracting is a very common procurement strategy that makes the application of unit pricing contracts possible. The structure, date, and timeline of payment are initially agreed in the conditions of the contract and articles of agreement before the construction project commences. This contractual arrangement shares some of the risks between the employer and the contractor. Using the cases of management contracting, continuous cost improvement strategies will be evaluated under this contractual arrangement.

a A management contractor working on a waste management plant is making use of multiple work package contractors. The unit pricing approach was preferred by a contractor who will ensure that the work package contractors will deliver their phased construction projects concurrently and according to the schedule. The management contractor is trying to innovate the delivery of the construction projects using the material life cycle costing approach and cost-benefit analysis. Do you think these continuous cost improvement strategies are relevant under this contractual arrangement? Do you think management contracting and unit pricing contracts make it difficult for the contractor to innovate?

_____

_____

_____

_____

b After several successful work package completions and unit pricing payments, the management contractor has identified an error in the in-house quantity surveyor's valuations. The quantity surveyor has computed the wrong labour costs and plant hire costs. This is the contractor's fault, and the employer has made the payments based on the agreed valuations for several months. Unit pricing contracts may be very beneficial to the client and contractor. In this situation, the contractor's cost forecasts and cash flow are looking favourable. How can the contractor use continuous cost improvement measures to reduce errors in the valuations of the work packages?

_____

_____

_____

_____

15.8 Cases on target cost contracts

A target pricing contract is an integrated contractual arrangement that ensures early involvement of the contractor, gain and pain sharing (Molenaar, Triplett, Porter, DeWitt & Yakowenko, 2007; Li, Wang, Yin, Kull & Choi, 2012). Target pricing has been compared with cost-reimbursable contracts because of the features. The consultant team works along with the contractor to identify cost-saving measures in the project. Hence, the option of target value design and continuous cost improvement measures such as material selection, specifications, market research, and circular economy can be integrated into a target costing contract. This commences from the planning phase, preferably after the briefing or conceptual design stages of the pre-contract phase of a construction project. The target cost contract provides an opportunity for cost savings and innovations in design, cost, constructability assessment, and waste reduction strategies. In the NEC4 contract document, target costing is an option for traditional sequential, novated design and build, and competitive design, and build procurement strategies.

a A government employer intends to construct a games village in a city hosting the next Olympics. The property department has recommended the target costing contract whereby the contractor will be involved very early in design and cost formations to reduce construction costs. The contractor's team conducted market research along with the cost planning team. Constructability assessment was also made after the design team confirmed that the cost limit was defined for the multiple building projects and facilities. The activities of the property department and the contractor seems to be continuous cost improvement. Do you agree with this assertion? How else can the team and contractor adopt more continuous cost-improvement strategies to drive down the construction cost and share the profits from the project?

_____

_____

_____

_____

b Under this same construction project, the contractor has the cost targets for each building element. However, owing to inflation and price fluctuations of construction materials such as steel, bricks, and timber, the cost targets are unreal. The contractor intends to adopt a continuous cost improvement strategy. Which of the strategies must be adopted to ensure that the construction project meets the target expected of the contractor?

_____

_____

_____

_____

15.9 Summary

Continuous cost improvement in a fixed-price contract is easier to adopt when compared with a lump sum contract because of the flexibility and risk-sharing arrangements. Remeasurement contracts may be tricky for the contractor to innovate if the valuation is not properly conducted. However, a direct design and build procurement arrangement will allow the contractors to implement their ideas freely. Cost reimbursable contracts are also very flexible; however, the contractor must inform the employer of changes to the scope or construction method. Innovative approaches such as continuous cost improvement can preferably be included in the contractual agreement before project commencement. Guaranteed maximum price contractors are similar to fixed-price contracts, and the impact for continuous cost improvement is even more severe. Unit pricing contracts do not permit much innovation under a management contract. Unit pricing contracts may be a stifling arrangement for continuous cost improvement strategies if the work package contractors are unwilling to accept new concepts on their projects. Moreover, the goal is to deliver their work packages very early under a fast-track arrangement. Target pricing contracts are the best arrangement for continuous cost improvement because the features are collaborative, accept innovative ideas and incorporate nearly all continuous cost improvement strategies.

References

Boukendour, S. and Bah, R. (2001). The guaranteed maximum price contract as call option. *Construction Management and Economics*, 19(6), pp. 563–567. doi: 10.1080/01446190110049848

Davis, P., Love, P., and Baccarini, D. (June 2008). Building procurement methods. *CRC Construction Innovation*, pp. 8–10. Available at: http://eprints.qut.edu.au/26844/1/26844.pdf

Davis, P. R. and Stevenson, D. (2004). Understanding and applying guaranteed maximum price contracts in Western Australia. In *Proceedings of the Australian Institute of Project Management 2004 National Conference.*, Perth, WA: Australian Institute of Project Management.

Li, H., Wang, Y., Yin, R., Kull, T. J., and Choi, T. Y. (2012). Target pricing: Demand-side versus supply-side approaches. *International Journal of Production Economics*, 136(1), pp. 172–184.

Miranda, M. L., Bauer, S. D., Aldy, J. E., and Podolsky, M. J. (1996). Unit pricing programs for residential municipal solid waste: an assessment of the literature. *Report prepared for US Environmental Protection Agency*, Washington DC, 40.

Molenaar, K. R., Triplett, J. E., Porter, J. C., DeWitt, S. D., and Yakowenko, G. (2007). Early contractor involvement and target pricing in US and UK highways. *Transportation Research Record*, 2040(1), pp. 3–10.

Murahashi, S-I. (1987). Contractual arrangements: A construction industry cost-effectiveness project report. *Tetrahedron Letters*, 28(21), pp. 2383–2386.

Nadel, N. A. (1991). Unit pricing and unbalanced bids. *Civil Engineering*, 61(6).

Nkuah, M. Y. (2006). Progress and performance control of a cost reimbursable construction contract. *Cost Engineering*, 4(5), pp. 13–18.

Picornell, M., Pellicer, E., Torres-Machi, C. and Sutrisna, M. (2017). Implementation of earned value management in unit-price payment contracts. *Journal of Management in Engineering*, 33(3), p. 06016001.

Zainordin, Z. M., Abd Rahman, N. A., Sahamir, S. R., and Khalid, Z. K. M. (2019). Methods of Valuing Construction Variation in Lump Sum Contract from the Public Client's Perspective. In *MATEC Web of Conferences* (Vol. 266, p. 03023). EDP Sciences.

SECTION D
DECISION TOOLS

16 Decision Tools for Continuous Cost Improvement – Part I

Temitope Omotayo, Udayangani Kulatunga, and Bankole Awuzie

List of Abbreviations

AND   Activity network diagram
BPMN   Business process modelling and notation
CMM   Capability maturity modelling
IDs   Interrelationship diagraphs
IDEF0   Icam DEFinition Function Zero
PDPCs   Process decision program charts

16.1 Introduction

Implementing continuous cost improvement in the construction process, irrespective of the nature, scale, and type of construction project, can benefit from several decision tools. The following tools in Sections 16.2–16.7 are discussed and illustrated with hypothetical projects as examples. The processes involved in attaining continuous cost improvement in the construction process is still hinged on the capacity of the project organisation to keep to a cost limit or benchmark, satisfy all stakeholders, and learn from previous mistakes. Likewise, as illustrated in Figure 16.1, the construction process indicates the points where continuous cost improvement can be infused into the construction project lifecycle using the traditional sequential procurement strategy.

As indicated in Figure 16.1, the first stage of continuous cost improvement uses documents from the inception briefing stage and feasibility studies. The first stage must determine whether the employer intends to use continuous cost improvement strategies, cost limits, and targets. This decision will ensure that the professional design and cost planning team will include continuous cost improvement measures in their plans and cost targets in stage 2, which will initially use the scheme design. Further improvements on the cost draw information from the product information and bill of quantities to create standardised systems for awarding contract and construction processes. In the final stage, mitigating waste measures must continually reduce the cost associated with the production phase.

DOI: 10.1201/9781003176077-20

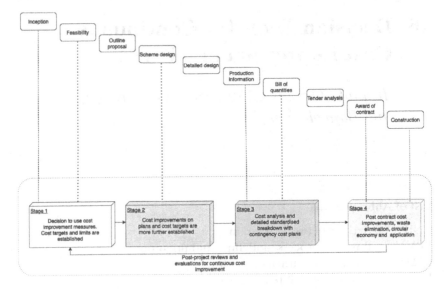

Figure 16.1 Continuous cost improvement stages in the project lifecycle.

Source: Authors.

Other continuous cost improvement strategies such as a circular economy that ensures reuse and waste minimisation on the site will aid improvements in construction costs. The process does not end at this stage because further improvement processes must benefit from evaluations in the post-project review phase for future projects.

Several tools can facilitate continuous cost improvement in the construction process. Some of these tools are as follows:

- Affinity diagrams
- Interrelationship diagraphs (IDs)
- Tree diagram
- Process decision programme charts (PDPCs)
- Matrix diagrams
- Prioritisation matrices

Each of the decision-making tools can be used at every continuous cost improvement stage identified in Figure 16.1.

16.2 Affinity diagrams

The affinity diagramming technique can be used as an information gathering and brainstorming technique (Harboe, Minke, Ilea, & Huang, 2012; Judge & McCrickard, 2008). The difficulty with affinity diagrams is keeping the notes after their creation (Judge & McCrickard, 2008). Affinity diagramming can be used in large-scale qualitative data organisation and analysis (Algozzine & Haselden, 2003; Harboe et al., 2012). In the instance of continuous cost improvement, all the stages in Figure 16.1 can use an affinity diagram. Affinity diagrams can also be used to improve existing processes through stakeholder brainstorming and for quality improvement and training purposes (Algozzine & Haselden, 2003; Lucero, 2015). Islam (2005) explained that affinity diagrams can be used when there is a need to change traditional thinking. The phases of applying affinity diagramming are articulating the question, brainstorming, positioning and scrubbing of ideas, grouping the ideas, creating headers, and prioritising ideas in a group (Islam, 2005). Affinity diagrams are applicable in value management systems or value stream mapping. Regarding continuous cost improvement, affinity diagrams can be used at the feasibility, scheme design, production information, tender analysis, and construction stages, as illustrated in Figure 16.1.

Affinity diagrams, as indicated in Figure 16.2, identify themes such as challenges, options, and implementation. For instance, if a construction planning team is faced with the challenge of labour shortage, rising cost of construction materials, and delays in delivery, there are options available. The brainstorming session will come up with ideas in sticky notes. These ideas are the options. In Figure 16.2, the options are off-site construction

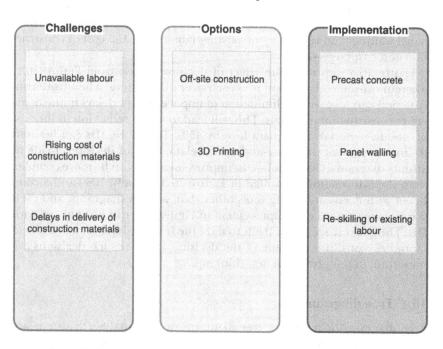

Figure 16.2 Affinity diagram for continuous cost improvement.

Source: Authors.

and 3D printing. The choice of off-site construction will lead to cost reductions by reducing labour using precast concrete, panel walling, and re-skilling of existing labour to fit the requirement of off-site construction. Generally, affinity diagramming for continuous cost improvement will aid the derivation of new strategies that can reduce construction costs and eliminate waste. With the use of off-site construction, waste will be eliminated, and reduced costs can be achieved.

16.3 Interrelationship diagraphs (IDs)

An interrelationship diagram (ID) is also known as a relations diagram. Alexander (2018) explained that IDs are one of the seven quality planning and management tools identified by Mizuno (1988). The application of IDs illustrates the causes and effects of decisions, factors, issues, or ideas (Winchip, 2001). IDs can be a useful tool in the affinity diagramming process when there is a need to provide an in-depth evaluation of the qualitative data (Winchip, 2001). The application of IDs also follows the phases of producing an affinity diagram. In applying IDs in continuous cost improvement, the feasibility development associated phase in Figure 16.2 can provide a clear example of how the impact of decisions can lead to further outcomes. Another example is Figure 16.2, where the implementation of new from the affinity diagram will seek to further explain what effects may be derived from adopting precast concrete, panel walling, and re-skilling of existing labour over the cost of construction and cost targets.

Figure 16.3 illustrates the effect of decisions made from Figure 16.2, wherein off-site construction is chosen over 3D printing. The offsite construction can lead to the elimination of importation or delays in the delivery of construction materials. This will lead to a 45% reduction in the cost of labour, materials, and plant hire by 45%. Therefore, IDs can be used to further present the cause-and-effect relationship of decisions made in affinity diagrams. Continuous cost improvement IDs can be more complex than the information presented in Figure 16.3. Equally, IDs can be combined with decision-making tools other than affinity diagrams, and each stage of continuous cost improvement in Figure 16.1 can be analysed with IDs. The general agenda of IDs is to show the relationship between two and more decisions, the outcome of the decision and why such decisions are important in the process of new thinking.

16.4 Tree diagrams

Tree diagrams or transition tree diagrams or tree analysis diagrams are used to itemise potential causes and their weighting factors that expressed the degree of feasibility (Hutchins, 2016; Jabrouni, Kamsu-Foguem, Geneste, & Vaysse, 2011). A tree diagram aims to identify the root cause

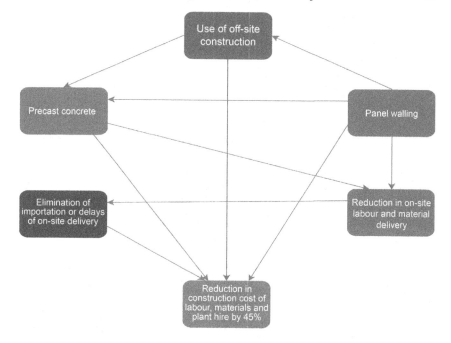

Figure 16.3 Interrelationship diagram for continuous cost improvement.

Source: Authors.

of a problem (Burch, Konevtsova, Heinrich, Höferlin, & Weiskopf, 2011). Jabrouni et al. (2011) acknowledged the four stages of producing a tree diagram as the following:

i Data collection: This can be done as a team for problem identification.
ii Causal factor charting: The question of why the problem occurred is asked.
iii Root cause identification: If the root cause cannot be identified, it will be documented.
iv Root cause validation: Revert to step (c) until the team agrees on identifying the root cause.

Tree diagrams can be applied as a tool in the continuous cost improvement process through the PDCA principle. The first step of the PDCA principle is the identification of problems. In this stage, the tree diagram can clearly identify a problem causing cost escalations in the construction phase. Creating the tree diagram is based on asking questions and investigating through brainstorming and problem identification using existing facts and documents. This may be conducted in a site meeting.

The root cause of cost escalation must be identified at the top or bottom of the tree analysis. Figure 16.4 illustrates a scenario where the root cause

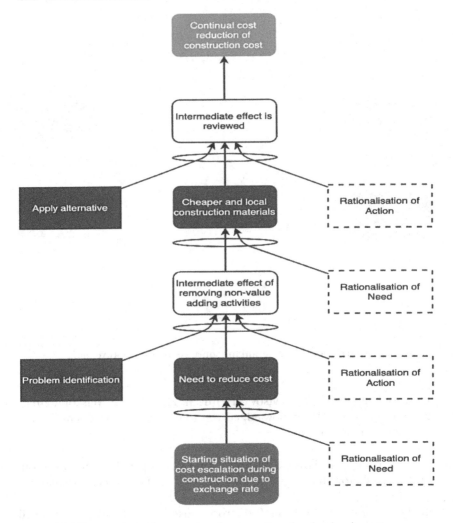

Figure 16.4 Tree analysis diagram showing a continual reduction in cost.

Source: Authors.

was identified at the bottom as a starting situation of cost escalation due to the unfavourable exchange rate. Further identification of problems noted the non-value adding activities are contributing to the rising cost of construction. The actions taking are identified on the left-hand side of Figure 16.4. On the right-hand side, the realisation of need and action are noted to ensure that the contractor or construction project management responsibility for the cost monitoring must act. The intermediate effects of the actions are documented and reviewed before the final goal of continual cost reduction is achieved at the top of the tree. In other tree analysis diagrams, the process can either start from the bottom upward or vice versa.

16.5 Process decision program charts (PDPCs)

Process decision program charts (PDPCs) are used as a methodological approach for debugging plans under development (Kochan, Gittell, & Lautsch, 1995; Mizuno & Bodek, 2020). PDPCs are used in quality improvement, management, and continuous improvement (Kovach, Cudney, & Elrod, 2011; Mizuno & Bodek, 2020). PDPCs, which are used as countermeasures, are created to ensure problems in plans are mitigated. When a complex plan is adopted in a project or completed on schedule, PDPCs can be used. In mega construction projects such as dams, rail, bridges, power plants, the cost and schedule overruns can be very expensive. In this situation, PDPCs can be used along with a tree analysis diagram to eliminate errors before implementation. The process of using PDPCs in continuous cost improvement is as follows:

i Create a high-level tree analysis diagram of the proposed construction. The objectives and levels of activities must be clearly defined. The high-level tree analysis diagram can have up to three levels.
ii Conduct brainstorming can be conducted on the tasks identified in the third level or final level.
iii Identify and eliminate all problems that may hinder the implementation of the plan. The problems can become the fourth or next level.
iv Conduct a brainstorming session on the problem to develop mitigating measures, illustrated in the fifth or next level. The outcomes of this task must document the solutions to the identified problems.
v Evaluate the mitigating measures to understand their feasibility within the project. This evaluation considers the measures in terms of schedule, budgetary, risk, quality, and resources implications.

Based on the continuous cost improvement tree analysis of Figure 16.4, the PDPCs for continuous cost improvement in Figure 16.5 consist of five levels. The third level provides the alternative continuous cost improvement strategies for the problems identified in level 2. The third level is assessed for the problems that may occur during implementation. These challenges are stated in the fourth level of the figure. Recommended solutions to challenges identified in level 4 are stated in the fifth level of the chart in Figure 16.5. Therefore, PDPCs are used to further analyse plans as illustrations, brainstorm the weaknesses and feasibility of the plan for implementation and end with a fine-tuned plan. Further refinements on the fifth level can be conducted to ensure the practicality of the plan.

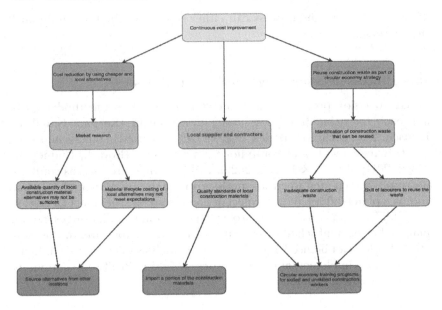

Figure 16.5 Process decision program chart for continuous cost improvement.

Source: Authors.

16.6 Matrix diagrams

Matrix diagrams are very common in presenting the outcomes of a plan and continuous improvement in a chart (Kovach, Cudney, & Elrod, 2011). A matrix diagram provides a two-dimensional table that enables the comparative presentation of issues (Van der Meij, Van Amelsvoort, & Anjewierden, 2017). Localisation can be used to organise the information in a matrix. This process distinguishes between topical and categorical localisation (Kauffman & Kiewra, 2010). Matrix diagrams can be used to identify relationships between variables or attributes in two-dimensional rows and columns (Miner, 2001). Matrix diagrams can also be used for collaborative learning, brainstorming sessions, and new thinking. When thoughts and ideas are collated through mind mapping, the categorisation of themes can be conducted and presented in a matrix diagram (see Figure 16.6).

Figure 16.6 provides examples of a simple matrix diagram identifying the causations of problems resulting in a divergent cost target in the execution phase of a construction project. The first perspective notes subcontractor's cost issues, rising costs associated with a preliminary item, and overheads. Another perspective collates the issues of delivery delays, import duties, and exchange rate fluctuations. Labour-related productivity and skills comprise perspective 3. The fourth perspective recognises plan hire cost, wastage on the construction site and delayed payments by the employer as some of

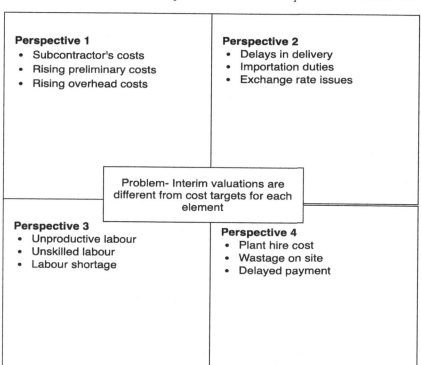

Figure 16.6 A simple matrix diagram identifying the root cause of issues in construction cost.

Source: Authors.

the issues resulting in the cost targets going beyond the benchmark. The purpose of matrix diagrams is to isolate the root cause of issues clearly for further analysis. The process of designing a matrix diagram can be achieved through a site meeting session. Site meetings have been discussed in Chapters 7, 8, and 9 as important in guaranteeing the categorisation of problems, which is the first step of the PDCA circle. Hence the matrix diagram can support the Deming cycle and further continuous cost improvement, especially in the construction process and interim evaluations.

16.7 Prioritisation matrices

Prioritisation matrices are used for multi-criteria decision-making (Bryson, 2018; Lussier & Liggett, 2004). While matrix diagrams are used to localise and categorise issues, prioritisation matrices are larger and more complex. Prioritisation matrices use weightings such as the Saaty's scale to determine which attribute is more important than the other (Islam, 2005; Omotayo,

(1 = Bad, 10 = Good)

Solutions	C1	C2	C3	C4	C5	Total
Solution A- Recycle waste	2	4	4	3	1	14
Solution B- Reuse waste	5	2	4	2	3	16
Solution C- Adopt off-site construction	3	4	4	1	2	14
Solution D- Seek extension of time	5	3	5	2	3	18
Solution E- Find local suppliers and plant hire	3	5	4	2	3	17
Solution F- Training in the workplace	3	1	4	5	3	16

Keys:
- C1: Easy to implement
- C2: Feasible
- C3: Cost effectiveness
- C4: Time
- C5: Easy to maintain

Figure 16.7 Solution selection matrix as an example of prioritisation matrix.

Source: Authors.

Awuzie, Egbelakin, Obi, & Ogunnusi, 2020; Hutchins, 2016; Yazdani & Tavakkoli-Moghaddam, 2012). In some instances, prioritisation matrices can present the outcome regarding the degree of impact, severity or difficulty from high to low in the same format as the matrix diagram of Figure 16.6. Another option for the prioritisation matrix is the solution selection matrix (see Figure 16.7).

Solution selection matrices are an example of a prioritisation matrix that can be used in a continuous cost improvement scenario where there is a cost overrun due to the rising cost of construction materials and importation delays. Figure 16.7 presents a situation where the continuous improvement manager intends to weigh the six solutions using a scale of 1 to 10 with five categories. Category C1 is the ease to implement, C2 is the feasibility, C3 is cost-effectiveness, C4 relates to time, and C5 is the ease of maintenance. In the solution matrix, Solution D suggests an extension of time for the importation of construction materials, and solution E which is the option to find local suppliers and plant hire companies. The application of prioritisation matrices in the continuous cost improvement process can be combined with other decision tools such as the tree analysis diagram or PDPCs.

16.8 Summary

Although there are numerous decision-making tools for continuous cost improvement for construction process improvement, a combination of one or more tools other than the PDCA chart can be applied in the construction process. Affinity and interrelationship diagrams can be applied in constructability assessment and brainstorming sessions to understand the implications of decisions. Construction cost management cannot stand alone without considering the programme of works, designs, bills of quantities, and contractual arrangements. Hence, the decision made in each tender document may be subjected to detailed evaluations with the use of tree diagrams, PDPCs, matrix diagrams, and prioritisation matrices. This second part of decision-making tools for continuous cost improvement in the construction industry is discussed in the next chapter.

References

Alexander, M. (2018). The interrelationship digraph using R. *Software Quality Professional*, 20(3), pp. 49–54.

Algozzine, B. and Haselden, P. G. (2003). Tips for teaching: Use of affinity diagrams as instructional tools in inclusive classrooms. *Preventing School Failure: Alternative Education for Children and Youth*, 47(4), pp. 187–189.

Bryson, C. (2018). Prioritization matrix use in program/project management. *Quality*, 57(9), p. 20.

Burch, M., Konevtsova, N., Heinrich, J., Höferlin, M., and Weiskopf, D. (2011). Evaluation of traditional, orthogonal, and radial tree diagrams by an eye tracking study. *IEEE Transactions on visualization and Computer Graphics*, 17(12), pp. 2440–2448.

Harboe, G., Minke, J., Ilea, I., and Huang, E. M. (2012). Computer support for collaborative data analysis: augmenting paper affinity diagrams. In *Proceedings of the ACM 2012 Conference on Computer Supported Cooperative Work*. pp. 1179–1182.

Hutchins, D. (2016). *Hoshin Kanri: The strategic approach to continuous improvement*. Routledge.

Islam, R. (2005). Rioritization of ideas in an affinity diagram by the AHP: An example of k-economy. *IIUM Journal of Economics and Management*, 13(1), pp. 1–21.

Jabrouni, H., Kamsu-Foguem, B., Geneste, L., and Vaysse, C. (2011). Continuous improvement through knowledge-guided analysis in experience feedback. *Engineering Applications of Artificial Intelligence*, 24(8), pp. 1419–1431.

Judge, T. K. and McCrickard, D. S. (2008). Collaborating on Affinity Diagrams Using Large Displays. TR-08-19, Virginia Tech.

Kauffman, D. F. and Kiewra, K. A. (2010). What makes a matrix so effective? An empirical test of the relative benefits of signaling, extraction, and localization. *Instructional Science*, 38(6), pp. 679–705.

Kochan, T. A., Gittell, J. H. and Lautsch, B. A. (1995). Total quality management and human resource systems: An international comparison. *International Journal of Human Resource Management*, 6(2), pp. 201–222.

Kovach, J., Cudney, E. and Elrod, C. (2011). The use of continuous improvement techniques: A survey-based study of current practices. *International Journal of Engineering Science and Technology*, 3, pp. 89–100. doi: 10.4314/ijest.v3i7.7S

Lucero, A. (2015). Using affinity diagrams to evaluate interactive prototypes. In *Proceedings of the IFIP Conference on Human-Computer Interaction*, Springer, pp. 231–248.

Lussier, E. and Liggett, H. (2004). Potential problems with prioritization matrices – A case study. *IIE Annual Conference and Exhibition 2004*, pp. 989–993.

Miner, A. S. (2001). Efficient solution of GSPNs using canonical matrix diagrams. In *Proceedings 9th International Workshop on Petri Nets and Performance Models*, IEEE, pp. 101–110.

Mizuno, S. and Bodek, N. (2020). *Management for quality improvement: The seven new QC tools*. Productivity Press.

Omotayo, T., Awuzie, B., Egbelakin, T., Obi, L., and Ogunnusi, M. (2020). AHP-systems thinking analyses for Kaizen costing implementation in the construction industry. *Buildings*, 10(12), p. 230.

Van der Meij, J., Van Amelsvoort, M., and Anjewierden, A. (2017). How design guides learning from matrix diagrams. *Instructional Science*, 45(6), pp. 751–767.

Winchip, S. M. (2001). Affinity and interrelationship digraph: A qualitative approach to identifying organizational issues in a graduate program. *College Student Journal. Project Innovation (Alabama)*, 35(2), p. 250.

Yazdani, A. A. and Tavakkoli-Moghaddam, R. (2012). Integration of the fish bone diagram, brainstorming, and AHP method for problem solving and decision making – a case study. *International Journal of Advanced Manufacturing Technology*, 63(5–8), pp. 651–657. doi: 10.1007/s00170-012-3916-7

17 Decision Tools for Continuous Cost Improvement – Part II

Temitope Omotayo, Udayangani Kulatunga, and Bankole Awuzie

17.1 Introduction

The first part of decision-making tools for continuous cost improvement was discussed in Chapter 16. This chapter is a continuation of decision-making options decision-makers may use to implement continuous cost improvement in their construction projects. Further decision tools that will be presented in this chapter are the following:

- Activity network diagram (AND)
- Cause and effect diagrams
- The 5S diagram
- Business process modelling and notation (BPMN)
- Icam DEFnition Function Zero (IDEF0)
- Capability maturity modelling (CMM)

17.2 Activity network diagrams (ANDs)

An AND is primarily used in project management planning to identify important tasks and dependencies visually (Jamnuch & Vatanawood, 2019). As a continuous cost improvement tool, resources can be associated with critical activities to identify tasks with time constraints that must be focussed on. ANDs may be produced with the Microsoft Office Project Professional software tool or any other online tool (Zhou, Zhang, & Jiang, 2007). In most cases, ANDs are produced after developing a Gantt chart or other programme of works identifying the critical schedule and baselines of a construction project. ANDs are also crucial in understanding the tasks that must be completed or concurrent activities and workload information (Kovach, Cudney, & Elrod, 2011). Soković, Jovanović, Krivokapić, and Vujović (2009) and Bessant, Caffyn, and Gallagher (2001) identified ANDs as quality control tools for continuous improvement. In terms of continuous cost improvement, activity diagrams do not necessarily act as an improvement tool but rather a visual presentation of activities that must be crushed or reduced to meet the budgetary expectations of a construction project.

DOI: 10.1201/9781003176077-21

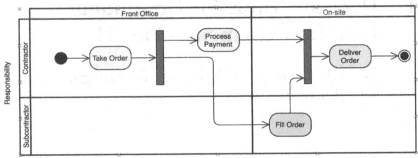

Figure 17.1 Activity network diagram showing material delivery between contractor and subcontractor.

Source: Authors.

Other than the common AND from MS Project Professional and other similar tools, an example in Figure 17.1 illustrates the dependencies for an order executed by a contractor from the front office for delivery to the construction site. In Figure 17.1, it is evident that the onsite delivery of the order depends on the subcontractor's responsibilities and the contractor's payment. The financial and schedule information can be improved by adding new tasks to ensure that the delivery process is less expensive, of the right quality, and on schedule.

17.3 Cause and effect

The cause and effect diagram, also known as the fishbone diagram, is used to present and evaluate the implications of decisions on a system (Aydin, Shaygan, & Dasdemir, 2014; Fukuda, 1978; Hossen, Ahmad, & Ali, 2017). Cause and effect diagrams have been used in quality improvement and control (Aydin *et al.*, 2014; Fukuda, 1978). The sort, set in order, shine, standardise, sustain (5S) methodology or 5 Whys cause and effect diagram may be developed to analyse decisions presented in a prioritisation or matrix diagram for continuous cost improvement. A cause and effect diagram can be used at any stage of continuous cost improvement. However, in the planning stage, where feasibilities studies are conducted, alternative construction methods and costs may be evaluated and presented with the aid of a cause and effect diagram. Thus cause and effect diagrams are useful tools for understanding the interrelationships between continuous cost improvement decisions and their final impact on the construction process. Figure 17.2 presents the post-project review of construction cost performance.

The effect of the problem, which is a cost overrun, was investigated through a post-project review evaluation. The causes are identified as

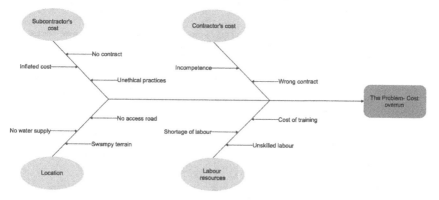

Figure 17.2 Cause and effect (fishbone) diagram illustrating the post-project review of a cost in a construction project.

Source: Authors.

location, labour resources, subcontractor's cost, and contractor's cost information. The details of each cause are presented in the arrow. The implications of using a cause and effect diagram are to expose the depths of the causations of understanding how they interact, leading to the final effect of cost overrun. Furtherance of this fishbone diagram can be the causal loop diagram under a systems-thinking approach. The cause and effect diagram will expose the causes of a problem for further analysis. Therefore, a fishbone diagram is a decision tool for analysis. The first step is problem identification in a PDCA circle.

17.4 The 5S diagram

The continuous improvement strategy of the 5S is important as a decision tool to provide a structured set of rules and tasks for enhancement (Sidhu, Kumar, & Bajaj, 2013; Singh & Singh, 2013). The 5S are sort, set in order, shine, standardise, and sustain as illuminated in Figure 17.3. A set of questions must be answered under each "S" category to attain continuous improvement (Dulhai, 2008; Mĺkva, Prajová, Yakimovich, Korshunov, & Tyurin, 2016).

Concerning continuous cost improvement, the questions are as follows:

- *Sort:* Is there is a logical order of tasks in the construction project? Are there plant, labour and material, and financial resources required to meet the needs of the individual tasks?
- *Set in order:* Are the tasks in the AND or programme of works set in the right order? Is there a systematised approach to payments, construction, and evaluation?
- *Shine:* Are the construction site and plans safe? Are the tasks in the programme of works waste-proof?

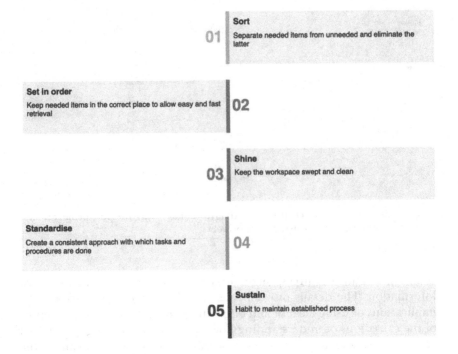

Figure 17.3 5S of continuous improvement.

Source: Authors.

- *Standardise:* Are the construction process and method statement clarified and known? Is there an established and practical work breakdown structure? Which methods are in place to eliminate non-value-adding activities?
- *Sustain:* Is there an established construction site behaviour to ensure the elimination of non-value adding activities? Is there is a quality and coordination culture aimed at eliminating unnecessary overhead costs?

The purpose of the 5S in continuous cost improvement is to serve as a guide and important decision-making tool in evaluating the readiness of the construction team for improvement activities on the site. In some instances, each of the questions asked in the 5S category may be scored on a scale of 1–5 to determine their availability and practicality.

17.5 Business process modelling and notation (BPMN)

Business process modelling and notation (BPMN) is a decision-making tool for illustrating and identifying tasks in a business with brainstorming to improve business models (Omotayo & Kulatunga, 2017b). Dijkman, Dumas, and Ouyang (2008) and Chinosi and Trombetta (2012) described BPMN as a de-facto standard for graphically showing each activity in an organisation.

Once a BPMN chart has been produced in a graphical structure as indicated in Figure 17.4a, the opportunities for improvement can be incorporated into the model. The BPMN in Figure 17.4a consists of the activities conducted by the employees, management staff, and supervisors. Each activity in Figure 17.4a contains nodes of problems, solutions, and improvement measures geared towards a continuous improvement process within the business organisation.

In most cases, the purpose of BPMN is to improve the financial position of an organisation. Therefore, BPMN is useful in continuously improving the overhead costs and creating new strategies in organisations. Figure 17.4b explains the business model of the manufacturer who acts as a

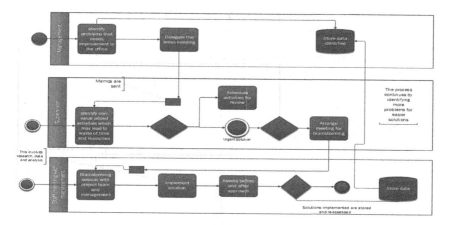

Figure 17.4a Business process modelling and notation (BPMN) for continuous cost improvement construction organisations.

Source: Authors.

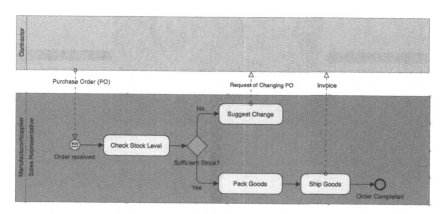

Figure 17.4b Business process modelling and notation for purchase order process improvement.

Source: Authors.

supplier for the contractor. Each of the activities in the business process of the manufacturer is represented in Figure 17.4b. The well-defined activity information in the BPMN of Figure 11.4b can be enhanced. For instance, the "sufficient stock" decision may always use the continuous improvement manufacturing system to keep the stock level ready for the contractor's purchase order. The BPMN has multiple applications in business process improvement. It may be applied in business strategy, manufacturing, payment systems, or communication systems. A good BPMN model will enable the management team to reduce non-value-adding activities and unnecessary overhead costs, thus creating continuous cost improvement.

17.6　Icam DEFinition Function Zero (IDEF0) for improving post-cost controlling activities

Icam DEFinition Function Zero (IDEF0) is a modelling tool for continuous improvement (Omotayo & Kulatunga, 2017a). IDEF0 is similar to the AND but uses an input, output, mechanism, and control as presented in Figure 17.5 (Kawai, Seki, Fuchino, & Naka, 2012; Presley & Liles, 1995). IDEF0 has a similar application to BPMN but provides a detailed approach to the tasks that must be improved. A standard IDEF0 will first provide a graphical view of the system and areas of improvement, and then the

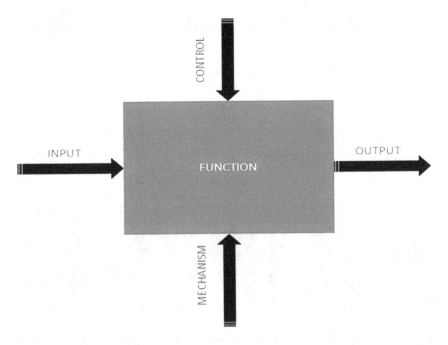

Figure 17.5 IDEF0 components.

Source: Authors.

improvement tasks will be infused into the first network of events. Soung-Hie and Ki-Jin (2000) described IDEF0 as the function in each activity carried out. The inputs are factors which the activities can alter. The control relates to external constraints that can impede the success of the activities, while the mechanism is the tool or means to fulfil the activity. The output is the result of the activity.

IDEF0 can be used to depict and decompose the contents of a work breakdown structure in a visual manner. This structure will represent a work breakdown of multiple layers. Therefore, IDEF0 may be very complex and difficult to view for a novice. Hence, there will be a parent IDEF0 that will be broken down into several children's IDEF0 models. For illustration, a simplified version of a continuous cost improvement model is presented using IDEF0.

17.7 Capability maturity model for continuous cost improvement

The capability maturity model (CMM) was developed by the Software Engineering Institute (SEI) for the software process improvement in the software industry (Keraminiyage, 2009). The practicality and application of CMM in the construction industry have some challenges which are mostly cost-related (Eadie, Perera, & Heaney, 2011; Keraminiyage, 2009). CMM can be used to measure the extent of implementing continuous improvement in an organisation, and when it comes to business improvement, it is a useful tool in creating a roadmap for optimising the adoption of continuous cost improvement (Omotayo *et al.*, 2019). Using the example of Figure 17.13, the CMM model was developed by (Omotayo *et al.*, 2019) to depict how the five levels of continuous improvement can help a construction organisation to improve their post-contract cost controlling activities on construction sites as well as improving their organisation.

Figure 17.6 explains the application of CMM in a construction organisation at various levels. CMM may be presented in a cross-function flow chart to identify the five levels as follows:

Level 1 – Initial or chaotic: This stage of construction is faced with an unstandardised process of construction, cost management, procurement, contract administration, selection, and award of contract. Most small- and medium-scale construction organisations are in this phase of construction characterised by multiple managerial issues and setbacks.

Level 2 – Repeated: This second level of CMM is also known as the tracked or planned phase. Construction organisations have learned lessons from their drawbacks and can create templates for their construction practice. Construction costs are planned with a digital tool, and there is a drive towards re-organisation and

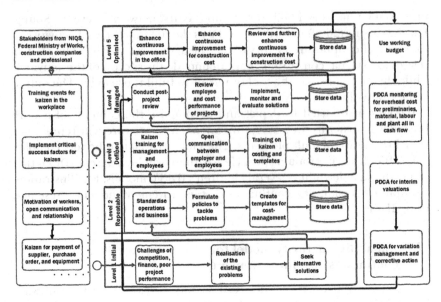

Figure 17.6 A capability maturity model for continuous improvement implementation.

Source: Authors.

defining the construction process. At this stage, continuous cost improvement will be discussed, and options for using digital tools may be considered.

Level 3 – Defined: The defined phase of CMM may also be described as good or best practice applications in construction. The construction processes are defined, and the need to improve will standardise tasks and subtasks. The desire to eliminate non-value-adding activities may be reviewed at this level. The essential construction cost management digital technology may be applied in this phase. Peer reviews and post-project reviews of employees and construction outputs will be more prominent in this CMM stage.

Level 4 – Managed: The managed phase of CMM is a quantitative level whereby extensive investments are placed on data storage, knowledge management, analysis, and constant post-project reviews to improve construction projects. The desire to satisfy clients and employees will be at the forefront of the management. The management level of CMM will also emphasise training and the development of a knowledge-based construction organisation. Construction cost management may use in-depth evaluations and improvement measures to reduce overhead costs and minimise physical and non-physical waste.

Level 5 – Optimised: The optimised level can be attained once the managed phase is developed to the point of constant improvement. Therefore, this is the continuous improvement level where all strategies can be deployed to train all employees and management. New developmental goals are set to ensure that cost information is stored and evaluated continually to increase the profitability of construction organisations.

17.8 Plan-do-check-act (PDCA) model

The plan-do-check-act (PDCA) model is the ultimate decision-making tool for continuous improvement. The application of the PDCA framework in construction cost management is not only applicable in the post-contract cost control phase in construction business strategy (Cooper & Slagmulder, 2017; Omotayo & Kulatunga, 2015; Omotayo. Awuzie, Egbelakin, Obi, and Ogunnusi, 2020; Ramezani & Razmeh, 2014). The applications of PDCA in the construction process are numerous. However, in terms of construction cost management, a holistic approach of micro-implementation is a better approach to complete implementation. Thus, the construction cost management lifecycle illustrated in Figure 17.1 provides a strong foundation for applying PDCA for cost improvement in the pre-tender or planning, analysis of tender and post-contract cost controlling phases.

An example of how PDCA may be useful in the post-contract cost control stage of a project is available in Figure 17.7. The "Plan" stage must identify

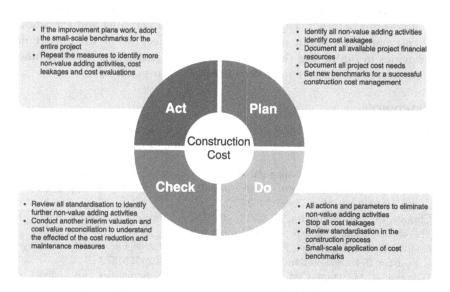

Figure 17.7 PDCA application for post-contract cost improvement.

Source: Authors.

all non-value-adding activities as the sources of the cost-related problems experienced in a construction project. All available financial resources and project cost needs must be reviewed and documented. New cost benchmarks must be provided as cost targets during construction. The "Do" phase of the model is the action phase, where a micro-implementation of the new cost reduction ideas is conducted. Standardisation measures are reviewed in this phase. The "Check" stage reviews the micro-implementation for its ability to reduce non-value adding activities and, more importantly, to meet the set benchmarks for the construction project. Finally, the "Act" phase considers the large-scale implementation of the improvement measures throughout the project and further identification of non-value-adding activities that may be eliminated.

17.9 Concluding remarks

Decision tools in continuous cost improvement can benefit from a combination of tools to examine and critique contract documents such as the bills of quantities, Gantt charts, and the programme of works. For instance, the programme of works containing the schedule and allocation of financial resources may be evaluated with the aid of PDPCs, tree analysis diagrams, and ANDs. The construction organisations can also use BPMN, IDEF0, and CMM to position their establishment strategically as a continuous improvement organisation. PDCA is the most widely used model and decision tool for continuous improvement. This chapter presented the model and how it can be applied to continually reduce construction costs. Continuous improvement must first be established in the organisation as a philosophy before it is applied in construction planning and execution of cost management. Site meetings and post-project reviews are important strategies in optimising the performance of organisations and attaining continuous cost improvement in construction processes.

References

Aydin, O. M., Shaygan, A. and Dasdemir, E. (2014). A cause and effect diagram and AHP based methodology for selection of quality improvement projects. *ENBIS-14*, September.

Bessant, J., Caffyn, S. and Gallagher, M. (2001). An evolutionary model of continuous improvement behaviour. *Technovation*, 21(2), pp. 67–77. doi: http://dx.doi.org/10.1016/S0166-4972(00)00023-7

Chinosi, M. and Trombetta, A. (2012). BPMN: An introduction to the standard. *Computer Standards & Interfaces*, 34(1), pp. 124–134.

Cooper, R. and Slagmulder, R. (2017). *Kaizen costing: Interorganizational cost management*, pp. 271–286. Portland: CRC. doi: 10.1201/9780203737873-13

Dijkman, R. M., Dumas, M. and Ouyang, C. (2008). Semantics and analysis of business process models in BPMN. *Information and Software Technology*, 50(12), pp. 1281–1294.

Dulhai, G. (2008). The 5S strategy for continuous improvement of the manufacturing processes in autocar exhaust. *Management and Marketing*, 3(4), pp. 115–120.

Eadie, R., Perera, S. and Heaney, G. (2011). Key process area mapping in the production of an e-capability maturity model for UK construction organisations. *Journal of Financial Management of Property and Construction*, 16(3), pp. 197–210. doi: 10.1108/13664381111179198

Fukuda, R. (1978). The reduction of quality defects by the application of a cause and effect diagram with the addition of cards. *International Journal of Production Research*, 16(4), pp. 305–319.

Hossen, J., Ahmad, N. and Ali, S. M. (2017). An application of Pareto analysis and cause-and-effect diagram (CED) to examine stoppage losses: A textile case from Bangladesh. *The Journal of the Textile Institute*, 108(11), pp. 2013–2020.

Jamnuch, R. and Vatanawood, W. (2019). Transforming activity network diagram with timed Petri nets. In *Proceedings of the 2019 12th International Conference on Information & Communication Technology and System (ICTS)*. IEEE, pp. 125–129.

Kawai, H., Seki, H., Fuchino, T., and Naka, Y. (2012). Pharmaceutical engineering strategy for quality informatics on the IDEF0 business process model. *Journal of Pharmaceutical Innovation*, 7(3–4), pp. 195–204.

Keraminiyage, K. P. (2009). *Achieving high process capability maturity in construction organisations*. Salford: University of Salford.

Kovach, J., Cudney, E. and Elrod, C. (2011). The use of continuous improvement techniques: A survey-based study of current practices. *International Journal of Engineering Science and Technology*, 3, pp. 89–100. doi: 10.4314/ijest.v3i7.7S

Mĺkva, M., Prajová, V., Yakimovich, B., Korshunov, A. and Tyurin, I. (2016). Standardization – one of the tools of continuous improvement. *Procedia Engineering*, 149, pp. 329–332.

Omotayo, T., Awuzie, B., Egbelakin, T., Obi, L. and Ogunnusi, M. (2020). AHP-systems thinking analyses for Kaizen costing implementation in the construction industry. *Buildings*, 10(12), p. 230.

Omotayo, T. S., Boateng, P., Osobajo, O., Oke, A. and Obi, L. I. (2019). Systems thinking and CMM for continuous improvement in the construction industry. *International Journal of Productivity and Performance Management*, 69(2), pp. 271–296. doi: 10.1108/IJPPM-11-2018-0417

Omotayo, T. and Kulatunga, U. (2015). The need for Kaizen costing in Indigenous Nigerian construction firms. University of Salford. *In Proceedings of the International Postgraduate Research Conference*, Media City, UK, 10–12 June 2015. doi: 10.13140/RG.2.1.3380.4405

Omotayo, T. and Kulatunga, U. (2017a). A continuous improvement framework using IDEF0 for post-contract cost control. *Journal of Construction Project Management and Innovation*, 7(1), pp. 1807–1823.

Omotayo, T. and Kulatunga, U. (2017b). A Gemba Kaizen model based on BPMN for small- and medium-scale construction businesses in Nigeria. *Journal of Construction Project Management and Innovation*, 7(1), pp. 1760–1778.

Presley, A. and Liles, D. H. (1995). The use of IDEF0 for the design and specification of methodologies. In *Proceedings of the 4th Industrial Engineering Research Conference*.

Ramezani, A. and Razmeh, A. (2014). Kaizen and Kaizen costing. *Academic Journal of Research in Business and Accounting*, 2, pp. 43–52.

Sidhu, B. S., Kumar, V. and Bajaj, A. (2013). The "5S" Strategy by using PDCA cycle for continuous improvement of the manufacturing processes in agriculture Industry. *International Journal of Research in Industrial Engineering*, 2(3), p. 10.

Singh, J. and Singh, H. (2013). Continuous improvement strategies: An overview. *IUP Journal of Operations Management*, 12(1), pp. 32–57.

Soung-Hie, Kim and Ki-Jin, Jang (2000). Designing performance analysis and IDEF0 for enterprise modeling in BPR. *International Journal of Production Economics*, 76(1), pp. 121–133.

Soković, M., Jovanović, J., Krivokapić, Z., and Vujović, A. (2009). Basic quality tools in continuous improvement process. *Journal of Mechanical Engineering*, 55(5), pp. 1–9.

Zhou, G., Zhang, G., and Jiang, P. (2007). Using extended activity-network diagram to design a process quality control model. In *Proceedings of the 2007 International Conference on Wireless Communications, Networking and Mobile Computing*, pp. 5111–5114.

Index

Printed in the United States
by Baker & Taylor Publisher Services